基于 MVC 和 EF 架构的监理信息系统开发实践

Development Practice of Supervision Information System Based on MVC and EF

王 新 著

北 京
冶金工业出版社
2015

内 容 提 要

本书以"建设工程监理信息系统"项目为例,重点介绍基于 MVC 和 EF 架构在 Web 模式下的监理信息系统的开发实践。全书分为七章,首先简要介绍了建设工程监理业务及工作要求,在此基础上运用系统分析和系统设计方法,归纳出建设工程监理信息系统的功能模型;然后系统介绍了监理信息系统的开发过程,包括在 VS2013 环境下建立系统结构、以 MVC 和 EF 架构进行业务流程设计和实体模型建立、系统功能导航的结构原理和实现方法、CRUD 模板及其自定义操作;最后全面介绍了系统功能的实现方法和过程。

本书内容注重实践,实例丰富,层次有序,结构鲜明。适合网站建设初学者、大中专学生、计算机软件开发人员阅读。

图书在版编目(CIP)数据

基于 MVC 和 EF 架构的监理信息系统开发实践/王新著. —北京:
冶金工业出版社,2015.7
ISBN 978-7-5024-6939-9

Ⅰ.①基… Ⅱ.①王… Ⅲ.①建筑工程—监理工作—管理信息系统—系统开发 Ⅳ.①TU712

中国版本图书馆 CIP 数据核字(2015)第 150722 号

出 版 人　谭学余

地　　址　北京市东城区嵩祝院北巷 39 号　邮编　100009　电话　(010)64027926
网　　址　www.cnmip.com.cn　电子信箱　yjcbs@cnmip.com.cn
责任编辑　杜婷婷　美术编辑　彭子赫　版式设计　孙跃红
责任校对　王永欣　责任印制　牛晓波

ISBN 978-7-5024-6939-9

冶金工业出版社出版发行;各地新华书店经销;固安华明印业有限公司印刷
2015 年 7 月第 1 版,2015 年 7 月第 1 次印刷
169mm×239mm;21.5 印张;419 千字;331 页
62.00 元

冶金工业出版社　投稿电话　(010)64027932　投稿信箱　tougao@cnmip.com.cn
冶金工业出版社营销中心　电话　(010)64044283　传真　(010)64027893
冶金书店　地址　北京市东四西大街 46 号(100010)　电话　(010)65289081(兼传真)
冶金工业出版社天猫旗舰店　yjgycbs.tmall.com

(本书如有印装质量问题,本社营销中心负责退换)

前　言

　　监理信息系统的设计、开发、应用是一个漫长的过程，但监理信息系统业务逻辑和数据处理算法并非想象之复杂，监理过程重在相关数据记录和文件流转及公文申报和审批，系统实现的关键在于实体模型及关系设计，系统开发的主要工作量来自数据视图建立和常规功能实现。根据以往项目建设和开发的经验，认为 MVC 和 EF 模式是不错的工具，因为其特有的建立在多层技术（包括模型、控制和视图）上的模板是快速建立数据处理控制器和 CRUD 操作的基础。

　　ASP. NET MVC 发展速度很快，从 3.0 版本更新到目前的 5.0 版本，不过几年时间。MVC 模式带来全新的开发者体验，ASP. NET 整合了模板应用系统，经由单一入口即可完成所有 Web 模板的选择和应用。EF 技术提供了实体模型定义和数据库管理的高度独立性，实现了"一次定义、多处应用"的优化模式，并且降低了管理信息系统开发过程中对数据库管理系统的依赖程度，配合 Linq 技术，为 MVC 的应用提供了强大的数据管理支持。ASP. NET 平台实现了 MVC 和 EF 技术的完美结合，辅助以衍生的工具，提高了 Web 项目建设开发的速度和运行的可靠性，为企业 Web 项目的建设开发提供强力的技术支持基础。当然，技术在发展，开发方法和方式在不断地变革，MVC 和 EF 技术对于 Web 项目的开发也在改进和发展中。

　　"建设工程监理信息系统"项目是在 Visual Studio 2013 集成环境中，利用 ASP. NET MVC 和 EF 相结合开发实现，是基于 Web 的管理信息系统。在项目分析阶段，根据用户需要，经过市场应用和开发技术的调研后，发现 ASP. NET MVC 在开发 Web 项目方面的优势，并对 Visual Studio 2010 集成开发平台的技术特点进行了深入的了解。在项目建设开发阶段，VS平台经历了多次升级，到2013版本时，已经将Web项目建设开发所涉及的技术，例如MVC、EF、jQuery、C#、HTML5、CSS3等集于一身。系统模板和实例库的使用平台为项目的快速建设提供了巨大的优势，也成就了本项目的实现和应用。

建设工程监理管理业务的特点表现在：业务区域广、时间周期长、参与部门多、工作责任大。如何依据行业管理标准加强科学管理、减少重复流程、共享监理成果、有效推进监理业务科学化、数据处理快速化和智能化、有力辅助决策，是建设工程监理部门和监理工程人员，特别是管理决策者所面临的管理效率瓶颈。

本书以"建设工程监理信息系统"项目为例，重点在于介绍基于MVC和EF架构在Web模式下企业管理信息系统的开发应用。现将其开发过程、方法和技术原理等内容整理成册，付诸出版，主要是基于以下几点：

（1）本项目的应用经过了实践的考验，稳定可靠，但仍有不断完善的必要。目前VS平台已升级至2015版本，新的方法和技术不断更新，因此，本项目还需要继续升级开发和完善。

（2）本项目的系统设计思想和数据模型设计方案被众多类似系统借鉴，充分体现"一个中心"向外辐射的"面向对象"的管理思想和思路，并且对大型企业管理信息系统建设同样有指导借鉴意义。

（3）在历年的"管理信息系统"及相关课程的教学实践中，本项目作为经典案例引入课堂，在培养学生理解和实践"管理信息系统"课程内容方面，作用显著；并且可以通过一个项目，同时传授多种技术应用方法。

（4）基于Visual Studio的ASP.NET MVC和相应的Web开发技术仍在持续发展，新技术、新方法不断完善，并得到推广应用，从而实现企业价值。

本书内容及编排具有以下特点：

（1）系统建设与开发综合了软件工程领域中的多种技术和方法，并通过管理信息系统学科，说明信息处理与信息系统，特别是管理信息系统的建设对推动企业或组织发展的重要作用。

（2）以实例为基础，以开发过程为线索，在说明系统建设内容的同时，说明相关开发技术和方法的运用实现，案例完整，系统性强。

（3）MVC 和 EF 是系统实现的基础和关键技术，实例多，内容逻辑关联多，注重方法实现原理，可以举一反三，融会贯通，具有普遍应用价值。

（4）系统功能设计实现是最后章节内容，便于系统的综合应用研究与集成调试。

（5）代码经过严格测试，排除了各种错误，包括因处理逻辑关系定义不全面而可能产生的"伪错误"。

在"建设工程监理信息系统"项目开发和本书编著过程中，得到了北京物资学院信息学院、项目实施相关单位和单位领导以及同事们的支持和帮助，本书由北京物资学院信息学院"智能物流系统北京市重点实验室 Beijing Key Laboratory（NO：BZ0211）"资助出版，在此一并表示感谢。

书中所述思想和方法是作者自己的实践经历，不足之处在所难免。在此，希望读者不吝指导，提出宝贵建议和意见。

作　者

2015 年 1 月

目 录

1 工程建设监理信息系统概述 ·· 1
  1.1 建设工程监理业务 ··· 1
    1.1.1 施工监理的前期准备工作 ································· 1
    1.1.2 施工准备阶段的监理 ····································· 3
    1.1.3 工程进度控制 ··· 5
    1.1.4 工程质量控制 ··· 7
    1.1.5 工程造价控制 ··· 9
    1.1.6 施工合同其他事项管理 ·································· 11
    1.1.7 其他监理工作 ·· 15
  1.2 系统建设内容 ·· 16
    1.2.1 监理业务逻辑分析 ······································ 16
    1.2.2 系统功能设计 ·· 20
    1.2.3 监理组织机构 ·· 27
    1.2.4 系统设计思想 ·· 29
    1.2.5 系统设计要求 ·· 30
  1.3 管理对象分析 ·· 31
    1.3.1 监理工程对象及属性 ···································· 31
    1.3.2 监理业务管理信息交换记录对象 ·························· 32
    1.3.3 系统服务对象 ·· 32
    1.3.4 辅助数据对象 ·· 33
  本章小结 ·· 34
  附录 ·· 34

2 建立工程监理信息系统项目 ·· 36
  2.1 Visual Studio 2013 简要概述 ····································· 36
    2.1.1 VS2013 的主要新功能 ·································· 36
    2.1.2 VS2013 开发环境 ······································ 37
    2.1.3 VS2013 新建项目 ······································ 39
    2.1.4 NuGet 程序包管理器 ··································· 42

2.1.5　引用目录内容 …………………………………… 46
　2.2　建立监理信息系统项目 ………………………………… 47
　　2.2.1　项目属性（Properties） ………………………… 47
　　2.2.2　区域目录（Areas） ……………………………… 50
　2.3　MVC目录架构 …………………………………………… 53
　　2.3.1　控制器目录（Controllers） ……………………… 53
　　2.3.2　模型目录（Models） …………………………… 53
　　2.3.3　视图目录（Views） ……………………………… 53
　　2.3.4　路由规则定义文件 ……………………………… 55
　　2.3.5　其他目录说明 …………………………………… 56
　2.4　Web.config文件 ………………………………………… 57
　　2.4.1　Web.config文件结构说明 ……………………… 57
　　2.4.2　主要节功能说明 ………………………………… 58
　　2.4.3　Web.config文件内容示例 ……………………… 60
　　2.4.4　本项目Web.config文件内容 …………………… 61
　本章小结 ………………………………………………………… 68

3　ASP.NET MVC架构及其应用 …………………………… 69
　3.1　ASP.NET MVC概述 …………………………………… 69
　　3.1.1　ASP.NET简介 …………………………………… 69
　　3.1.2　MVC设计模型 …………………………………… 71
　　3.1.3　MVC运行机制 …………………………………… 73
　　3.1.4　ASP.NET MVC的特点 …………………………… 75
　3.2　ASP.NET MVC项目的运行 …………………………… 76
　　3.2.1　路由规则定义 …………………………………… 76
　　3.2.2　路径命名与映射关系 …………………………… 78
　　3.2.3　布局页 …………………………………………… 80
　　3.2.4　_ViewStart.cshtml文件 ………………………… 85
　3.3　ActionResult与视图 …………………………………… 86
　　3.3.1　ActionResult的子类类型 ………………………… 86
　　3.3.2　ActionResult返回类型说明 ……………………… 87
　　3.3.3　View及其应用 …………………………………… 90
　3.4　Razor视图引擎 ………………………………………… 92
　　3.4.1　Razor标识符号 …………………………………… 92
　　3.4.2　Razor C#基本语法 ……………………………… 95

3.4.3　Razor C#循环语句 …………………………………………… 97
　　3.4.4　Razor C#判断语句 …………………………………………… 99
　　3.4.5　几个基于Razor帮助器的用法 ………………………………… 101
本章小结 …………………………………………………………………… 104

4　EF架构与实体模型定义 …………………………………………… 105
4.1　EF概述 ……………………………………………………………… 105
　　4.1.1　EF的特点 ……………………………………………………… 106
　　4.1.2　实体模型（EF）的验证规则 …………………………………… 106
　　4.1.3　EF Code First默认规则及配置 ………………………………… 108
4.2　A-工程管理实体模型定义 …………………………………………… 109
　　4.2.1　"工程信息"实体模型定义 …………………………………… 109
　　4.2.2　"工程图片"实体模型定义 …………………………………… 113
　　4.2.3　"工程增加"实体模型定义 …………………………………… 115
　　4.2.4　"单位工程"实体模型定义 …………………………………… 116
　　4.2.5　"工程调整"实体模型定义 …………………………………… 117
4.3　B-文档管理实体模型定义 …………………………………………… 118
　　4.3.1　"接收文件"实体模型定义 …………………………………… 118
　　4.3.2　"文件类别"实体模型定义 …………………………………… 119
　　4.3.3　"发出文件"实体模型定义 …………………………………… 120
　　4.3.4　"监理日记（个人登记）"实体模型定义 ……………………… 122
　　4.3.5　"监理日志（项目组登记）"实体模型定义 …………………… 124
4.4　K-系统管理实体模型定义 …………………………………………… 127
　　4.4.1　"系统用户"实体模型定义 …………………………………… 127
　　4.4.2　"用户增加"实体模型定义 …………………………………… 128
　　4.4.3　"用户登录"实体模型定义 …………………………………… 129
　　4.4.4　"系统角色"实体模型定义 …………………………………… 129
　　4.4.5　"系统功能"实体模型定义 …………………………………… 130
　　4.4.6　"角色功能"实体模型定义 …………………………………… 131
　　4.4.7　"用户登录日志"实体模型定义 ……………………………… 132
4.5　实体模型与数据库的关系 …………………………………………… 133
　　4.5.1　模型与DbContext类 ………………………………………… 133
　　4.5.2　psjldb12Context.cs类文件 …………………………………… 133
　　4.5.3　Web.config文件与<connectionStrings>节 ………………… 137
本章小结 …………………………………………………………………… 139

5 功能导航系统设计 ······ 140
5.1 系统功能管理 ······ 140
- 5.1.1 功能模块与子功能模块数据记录 ······ 140
- 5.1.2 系统功能管理控制器 ······ 141
- 5.1.3 功能数据记录列表显示视图 ······ 145
- 5.1.4 新增功能项目管理视图 ······ 147
- 5.1.5 功能项目详细内容显示视图 ······ 148
- 5.1.6 功能项目记录数据编辑视图 ······ 149
- 5.1.7 功能项目记录删除功能视图 ······ 150

5.2 系统角色管理 ······ 152
- 5.2.1 系统角色管理控制器 ······ 152
- 5.2.2 角色数据记录列表显示视图 ······ 155
- 5.2.3 新增角色功能视图 ······ 158
- 5.2.4 角色数据记录详细内容显示视图 ······ 159
- 5.2.5 角色数据记录编辑功能视图 ······ 160
- 5.2.6 角色记录删除功能视图 ······ 161

5.3 用户角色分配 ······ 162
- 5.3.1 一对多关系定义 ······ 162
- 5.3.2 系统角色记录检索 ······ 163
- 5.3.3 视图中实现用户角色选择 ······ 163

5.4 角色功能分配 ······ 164
- 5.4.1 系统角色实体与系统功能实体的关系 ······ 164
- 5.4.2 角色—功能分配功能实现的控制器 ······ 165
- 5.4.3 系统角色记录显示视图 ······ 167
- 5.4.4 功能记录显示的局部视图 ······ 169

5.5 用户登录与动态功能导航实现 ······ 170
- 5.5.1 系统用户登录方法 ······ 170
- 5.5.2 系统用户登录视图 ······ 172
- 5.5.3 系统主（一级）功能导航 ······ 174
- 5.5.4 子功能导航实现 ······ 176
- 5.5.5 子功能导航内容显示的局部视图 ······ 177

本章小结 ······ 178

6 CRUD 模板设计 ······ 179
6.1 CRUD 控制器模板应用实例 ······ 179
- 6.1.1 实体模型与数据库表的对应关系 ······ 180

- 6.1.2 建立 CRUD 控制器 …… 181
- 6.1.3 CRUD 控制器代码内容组成 …… 182
- 6.1.4 记录数据检索方法 …… 186
- 6.1.5 记录详细内容显示方法 …… 186
- 6.1.6 新增记录方法 …… 187
- 6.1.7 记录数据编辑方法 …… 188
- 6.1.8 记录删除方法 …… 189
- 6.2 CRUD 视图模板应用实例 …… 189
  - 6.2.1 记录列表显示视图 …… 190
  - 6.2.2 记录新增显示视图 …… 192
  - 6.2.3 记录详细内容显示视图 …… 194
  - 6.2.4 记录编辑显示视图 …… 195
  - 6.2.5 记录删除显示视图 …… 196
- 6.3 ASP.NET MVC 系统自有 CRUD 模板 …… 198
  - 6.3.1 控制器生成模板 …… 198
  - 6.3.2 记录列表显示视图生成模板 …… 204
  - 6.3.3 新增记录显示视图生成模板 …… 211
  - 6.3.4 记录详细内容显示视图生成模板 …… 219
  - 6.3.5 记录编辑视图生成模板 …… 225
  - 6.3.6 记录删除视图生成模板 …… 233
- 6.4 自定义 CRUD 模板 …… 240
  - 6.4.1 自定义控制器模板 …… 241
  - 6.4.2 自定义记录列表显示视图模板 …… 248
  - 6.4.3 自定义新增记录显示视图模板 …… 251
  - 6.4.4 自定义记录详细内容显示视图模板 …… 255
  - 6.4.5 自定义记录编辑显示视图模板 …… 258
  - 6.4.6 自定义记录删除显示视图模板 …… 262
- 本章小结 …… 265

7 系统功能设计与实现 …… 266
- 7.1 系统主页功能导航 …… 266
  - 7.1.1 主页内容组成结构 …… 266
  - 7.1.2 主页代码内容 …… 267
  - 7.1.3 代码功能说明 …… 270
  - 7.1.4 @RenderBody()方法的实现 …… 274

7.2 通用功能导航链接 ………………………………………………………… 275
7.2.1 用户切换 ……………………………………………………………… 275
7.2.2 工程选择 ……………………………………………………………… 279
7.2.3 用户注销 ……………………………………………………………… 284
7.2.4 修改密码 ……………………………………………………………… 285
7.2.5 系统主页 ……………………………………………………………… 288
7.2.6 关于我们 ……………………………………………………………… 291
7.3 工程管理功能实现 ……………………………………………………… 292
7.3.1 工程信息编辑 ………………………………………………………… 292
7.3.2 工程项目调整 ………………………………………………………… 300
7.3.3 工程分项管理 ………………………………………………………… 303
7.3.4 增加新的工程 ………………………………………………………… 305
7.3.5 删除当前工程 ………………………………………………………… 310
7.4 其他功能实现 …………………………………………………………… 313
7.4.1 J-查询统计 …………………………………………………………… 313
7.4.2 K-系统管理 …………………………………………………………… 318
7.4.3 L-基础数据 …………………………………………………………… 322
7.4.4 O-其他辅助信息管理 ………………………………………………… 327
本章小结 ……………………………………………………………………… 330

参考文献 ……………………………………………………………………… 331

1 工程建设监理信息系统概述

"工程建设监理信息系统"以北京磐石建设监理有限责任公司的工程监理业务管理为基础,以北京市地方性标准《工程建设监理规程》(DBJ 01-41—2002)为参考标准,借助 Visual Studio 开发工具,利用 ASP. NET MVC 和 EF 系统架构,设计建设开发。本章主要内容如下:

1.1 建设工程监理业务
1.2 系统建设内容
1.3 管理对象分析

1.1 建设工程监理业务[1]

北京磐石建设监理有限责任公司(原北京市磐石市政建设监理公司),起始于1989年,并于1991年7月正式成立。该公司是国家建设部批准的首批具有市政工程甲级资质的建设监理单位,1999年增补建筑工程监理甲级资质而成为双甲级建设监理单位。公司可跨地区、跨部门承接市政基础设施工程和工业与民用建筑工程等专业项目监理服务,并具有国外承揽项目管理的业务能力和实践。建立适合建设工程监理业务管理的智能信息化系统,有利于提升监理业务管理质量,扩展公司业务范围;同时有利于科学管理,提高效率。

根据北京市《工程建设监理规程》要求,工程建设监理的任务分为施工监理的前期准备、施工准备阶段的监理、工程进度控制、工程质量控制、工程造价控制、施工合同管理、施工安全管理、施工风险管理和整体协调等,可概况为"三控制、三管理和一协调"。

1.1.1 施工监理的前期准备工作

施工监理的前期准备工作主要由监理单位主持完成,其工作任务包括:
(1)建立工程项目监理机构。监理单位根据相关的监理合同,参考相关的

---

[1] 本节内容参考北京市地方性标准《建设工程监理规程》(DBJ 01-41—2002)编写。

监理规范及法律法规，建立相应的工程项目监理机构——项目监理部，并配备所需要的监理人员，根据相应的职责分配任务，之后报告工程建设单位、工程施工单位以及其他相关单位。

监理机构中的岗位有：总监理工程师、总监理工程师代表、监理工程师和监理员等，每个岗位上的工作人员根据岗位职责，负责相应的工作和任务。

（2）准备监理设施与设备。根据监理任务和监理工作的要求与需要，项目监理部向建设单位、施工单位、自己所在的监理单位申请必要的设施和设备，以备监理工作所用，监理工作完成后，如实归还。

（3）核查审阅施工图纸。总监理工程师收集工程施工图纸，存档管理。组织项目监理部工作人员阅读，了解工程特点及施工要求。在此基础上，进行全面审查，对于发现的问题、提出的建议，汇总整理，并提交相关单位。

（4）核查审阅相关合同。合同分为"委托监理合同"和"建设工程施工合同"，要求收集整理，分别审阅检查。审阅检查内容及要求如图1-1所示。

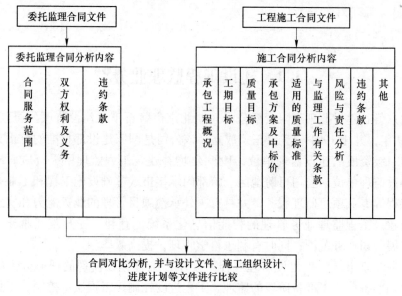

图1-1 合同文件检查内容

（5）编制工程项目监理规划和监理实施细则。在上述工作的基础上，总监理工程师组织项目监理部工作人员，制定监理规划和监理实施细则。

监理规划包括内容如下：

1) 工程项目概况：工程项目名称、建设地点、建设规模、工程类型、工程特点、参与单位名录（建设单位、设计单位、承包单位、主要分包单位）等；

2) 监理工作依据：政策法规；

3) 监理范围和目标：监理工作范围及工作内容、工期控制目标、工期质量控制目标、工程造价控制目标等；

4）工程进度控制：工期控制目标的分解、进度控制程序、进度控制要点、控制进度风险指标等；

5）工程质量控制：质量控制指标的分解、质量控制程序、质量控制要点、控制质量风险的措施等；

6）工程造价控制：造价控制指标的分解、造价控制程序、造价控制要点、控制造价风险的措施等；

7）合同及其他事项管理：工程变更、索赔管理、管理程序、合同争议协调方法等方面的要点及说明；

8）项目监理部情况说明：组织形式、人员情况、职责分工、人员进场计划安排等；

9）监理工作管理制度：信息和资料管理制度、监理工作报告制度、其他监理工作制度。

监理实施细则包括的内容有：专业工程的特点、监理工作的流程、监理工作的控制要点及目标值、监理工作的方法及措施等。

在监理工作实施过程中，监理实施细则应根据实际情况进行补充、修改和完善。

1.1.2 施工准备阶段的监理

施工准备阶段的监理工作内容主要包括：参与设计交底、审核施工组织设计（施工方案）、查验施工测量放线成果、组织第一次工地会议、施工监理交底、核查开工条件。

1.1.2.1 参与设计交底

首先，设计交底由建设单位主持，设计单位、承包单位、监理单位的项目负责人及相关人员参加。

监理工程师需要了解的内容有：建设单位对工程的要求；施工现场的自然条件（地形、地貌等），工程地质与水文地质条件；设计主导思想，建筑艺术要求与构思，使用的设计规范，抗震烈度和等级，基础设计，主体结构设计，装修设计，设备设计（设备造型）；对基础、结构及装修施工的要求，对建材的要求，对使用新技术、新工艺、新材料的要求，对施工过程中特别注意事项的说明；设计单位对监理单位和承包单位的施工图纸所提出问题的答复。

项目监理部需认真记录，承包单位负责收集整理（监理部设计格式）。所有有关工程变更要求的内容，必须经过建设单位、设计单位、监理单位和承包单位签认。

1.1.2.2 审核施工组织设计（施工方案）

施工组织设计（施工方案）的审核程序如下：

首先，承包单位在开工前向项目监理部报送施工组织设计（施工方案），同时填写《工程技术文件报审表》（格式及内容请参见《工程建设监理规程》（DBJ 01-42—2002）中附录 A-A1，以下同）；第二，总监理工程师组织审查并核准，需要修改的，由总监理工程师签署意见并退回承包单位，修改后再报，重新审核；第三，对于重大工程，项目监理部还需要报监理单位技术负责人审核后，再由总监理工程师签认，发还承包单位；第四，施工过程中，承包单位需要修改，仍需要报总监理工程师审核同意；第五，规模较大、工艺较复杂、群体或分期出图的工程，可分阶段报批施工组织设计；第六，技术复杂或采用新技术的分项、分部工程，承包单位需要编制相应的施工方案，报送项目监理部审核。

施工组织设计（施工方案）的审核内容如下：

承包单位的审批手续是否齐全、有效；施工总平面布置示意图是否合理；施工布置是否合理、施工方案是否可行、质量保证措施是否可靠并具有针对性；工期安排是否满足建设工程施工合同要求；进度计划是否能保证施工的连续性和均衡性，所需的人力、材料、设备的配置与进度计划是否协调；承包单位项目经理部的质量管理体系、技术管理体系和质量保证体系是否健全；安全、环保、消防和文明施工措施是否符合有关规定；季节施工方案和专项施工方案的可行性、合理性和先进性；总监理工程师认为应审核的其他内容。

审核施工组织设计（施工方案）的主要任务是审核《工程技术文件报审表》。

1.1.2.3 查验施工测量放线成果

查验施工测量放线成果的内容有：承包单位专职测量人员的岗位证书及测量设备检定证书；承包单位填写的《施工测量放射线报验表》（附录 A-A2），记录施工测量方案、红线桩的校准成果、水准点的引用测量成果、施工现场设置平面坐标控制网（或控制导线）、调和控制网；相应的保护措施是否有效。

1.1.2.4 第一次工地会议

参会人员：建设单位驻现场代表及有关职能人员；承包单位项目经理部经理及有关职能人员；分包单位主要负责人；监理单位项目监理部总监理工程师及全体监理人员。

会议内容：建设单位负责人宣布项目总监理工程师并授权；建设单位负责人宣布承包单位及其驻现场代表（项目经理部经理）；建设单位驻现场代表、总监理工程师和项目部经理相互介绍各方组织机构、人员及其专业、职务分工；项目经理部汇报施工现场的准备情况；会议各方确定协调方式、参加例会的人员、时间及安排；其他事项。

第一次工地会议后，由监理单位负责整理编印会议纪要，并分发给有关各方。

1.1.2.5 施工监理交底

施工监理交底由总监理工程师主持,参加人员有承包单位项目经理及有关职能人员、分包单位主要负责人、项目监理部监理工程师及监理人员。主要内容包括:明确适用的有关工程建设监理的政策、法令、法规等;阐明有关合同中约定的建设单位、监理单位和承包单位的权利和义务;介绍监理工作内容;介绍监理工作的基本程序和方法;提出有关报表的报审要求及工程资料的管理要求。

会后,由项目监理部编写会议纪要,并发给承包单位。

1.1.2.6 核查开工条件

核查开工条件的工作程序及内容如下:

首先,承包单位认为达到开工条件时应向项目监理部申报《工程动工报审表》(附录A-A5)。第二,监理工程师进行检查,检查内容有:政府主管部门已签发的"××市建设工程开工证";施工组织设计已经由项目总监理工程师审核;测量控制桩已查验合格;承包单位项目经理部人员已到位,施工人员、施工设备已按计划进场,主要材料供应已落实;施工现场道路、水、电、通讯等已达到开工条件。第三,监理工程师审核认为具备开工条件时,由总监理工程师在承包单位报送的《工程动工报审表》上签署意见,并报送建设单位。

1.1.3 工程进度控制

工程进度控制依据建设工程施工合同约定工期目标,在确保工程质量和安全并符合控制工程造价的原则下,采用动态的控制方法,对工程进行主动控制。

1.1.3.1 工程进度控制的基本程序

工程进度控制的基本程序如图1-2所示。

1.1.3.2 工程进度控制的内容和方法

审批进度计划的内容和方法有:承包单位根据建设工程施工合同约定,按时编制施工总进度计划、季度进度计划和月度进度计划,并按时填写《施工进度计划报审表》(附录A-A3),报项目监理部审批;监理工程师根据工程的条件(工程的规模、质量标准、复杂程度、施工的现场条件等)及施工队伍的条件,全面分析承包单位编制的施工总进度计划的合理性和可行性;施工总进度计划应符合施工合同中竣工日期规定,可以用横道图或网络图表示,并附有文字说明,监理工程师对网络计划的关键路线进行审查、分析;对于季度及年度进度计划,应要求承包单位同时编写主要工程材料、设备的采购及进场时间等计划安排;项目监理部应对进度目标进行风险分析,制定防范性对策,确定进度控制方案;总进度计划经总监理工程师批准实施,并报送建设单位,如需要重新修改,应限时要求承包单位重新申报。

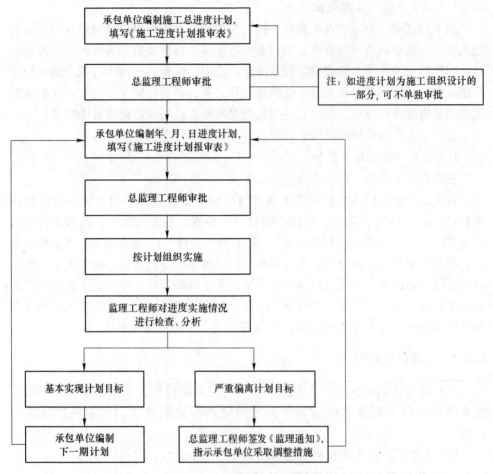

图1-2 工程进度控制的基本程序

进度计划的实施监督：项目监理部应依据总进度计划，对承包单位实际进度进行跟踪监督检查，实施动态控制；应按月检查月实际进度，并将其与月计划进度比较的结果进行分析、评价，如发现偏离应签发《监理通知》（附录B-B1），要求承包单位及时采取措施，实现计划进度目标；要求承包单位每月25日前报《（ ）月工、料、机动态表》（附录A-A9）。

工程进度计划的调整：发现工程进度严重偏离计划时，总监理工程师应组织监理工程师进行原因分析，召开各方协调会议，研究应采取的措施，并应指令承包单位采取相应调整措施，保证合同约定的目标实现。总监理工程师应在监理月报中向建设单位报告工程进度和所采取的控制措施的执行情况，提出合理预防由建设单位原因导致的工程延期及其相关费用索赔的建议；必须延长工期时，应要求承包单位填报《工程延期申请表》（附录A-A15），报项目监理部；总监理工程师依据施工合同约定，与建设单位共同签署《工程延期审批表》，要求承包单位据此重新调整工程进度计划。

1.1.4 工程质量控制

工程质量控制的内容包括：以工程施工质量验收统一标准及验收规范等为依据，督促承包单位全面实现施工合同约定的质量目标；对工程项目施工全过程实施质量控制，以质量预控为重点；对工程项目的人、机、料、法、环等因素进行全面的质量控制，监督承包单位的质量管理体系、技术管理体系和质量保证体系落实到位；严格要求承包单位执行有关材料、施工试验制度和设备检验制度；严禁不合格的建筑材料、构配件和设备在工程上使用；坚持本工序质量不合格或未进行验收不予签认，下一道工序不得施工。

工程质量控制主要包括：事前控制、事中控制、竣工验收、问题和事故处理等。

1.1.4.1 工程质量的事前控制

事前控制的主要工作内容包括：核查承包单位的质量管理体系，包括承包单位的机构设置、人员配备、职责与分工的落实情况；各级专职质量检查人员的配备情况；各级管理人员及专业操作人员的持证情况；检查承包单位质量管理制度是否健全。审查分包单位和试验室的资质，承包单位填写《分包单位资质报审表》（附录A-A6）；核查分包单位的营业执照、企业资质等级证书、专业许可证、岗位证书等；核查分包单位的业绩；经审查合格，签批《分包单位资质报审表》。查验承包单位的测量放线，应要求承包单位填写《施工测量放线报验表》，并附楼层放线记录。签认材料的报验，要求承包单位应按有关规定对主要原材料进行复试，并将复试结果及材料备案资料、出厂质量证明等附在《工程物资进场报验表》（附录A-A4）上。签认建筑构配件和设备报验，审查构配件和设备厂家的资质证明及产品合格证明、进口材料和设备商检证明，并要求承包单位按规定进行复试。应参与加工订货厂家的考察、评审，根据合同的约定参与订货合同的拟定和签约工作。合格后，填写《工程物资进场报验表》（附录A-A4）。检查进场的主要施工设备，要求承包单位在主要施工设备进场并调试合格后，填写《（ ）月工、料、机动态表》（附录A-A9）。审查主要分部（分项）工程施工方案，应要求承包单位对某些主要分部（分项）工程或重点部位、关键工序在施工前，将施工工艺、原材料使用、劳动力配置、质量保证措施等情况编写专项施工方案，并填写《工程技术文件报审表》。

1.1.4.2 施工过程中的质量控制

事中控制的程序为：首先应对施工现场有目的地进行巡视和旁站，如发现问题应及时要求承包单位予以纠正，并记入监理日志；第二，对所发现的问题可先口头通知承包单位改正，然后应及时签发《监理通知》，承包单位应将整改结果填写《监理通知回复单》，报监理工程师进行复查；第三，核查工程预检，要求承包单位填写预检工程检查记录，报送项目监理部核查；第四，验收隐蔽工程，要求承包单位按有关规定对隐蔽工程先进行自检，自检合格，将隐蔽工程检查记录报送项目监理部，然后由其对隐蔽工程检查记录的内容到现场进行检测、核查，对检测不合格的工程，应填写《不合格项处置记录》（附录B-B3），要求承包单位整改，合格后再予以复查；第五，分项工程验收，要求承包单位在一个检验批或分项工程完成并自检合格后，填写《分项/分部工程施工报验表》（附录A-A7），对报验的资料进行审查，并到施工现场进行抽检、核查，签认符合要求的分项工程，对不符合要求的分项工程，填写《不合格项处置记录》，要求承包单位整改；第六，分部工程验收，应要求承包单位在分部工程完成后，填报《分项/分部工程施工报验表》，总监理工程师根据已签认的分项工程质量验收结果签署验收意见。单位工程基础分部已完成、进入主体结构施工时，或主体结构完成、进入装修前应分别进行基础和主体工程验收，要求承包单位申报基础/主体工程验收；并由总监理工程师组织建设单位、承包单位和设计单位共同核查承包单位的施工技术资料，进行现场质量验收，并会同各方在基础/主体工程验收记录上签字认可。

1.1.4.3 工程竣工验收

（1）对发现影响竣工验收的问题签发《监理通知》，要求承包单位进行整改。

（2）总监理工程师组织竣工预验收，要求承包单位在工程项目自检合格并达到竣工验收条件时，填写《单位工程竣工预验收报验表》（附录A-A8），并附相应竣工资料（包括分包单位的竣工资料），申请竣工预验收。

（3）总监理工程师组织监理工程师和承包单位共同对工程进行检查验收，经验收需要对局部进行整改的，应在整改符合要求后再验收，直至符合合同要求，总监理工程师签署《单位工程竣工预验收报验表》。

（4）预验收合格后，监理单位应对工程提出质量评估报告，整理监理资料，工程质量评估报告必须经总监理工程师和监理单位技术负责人审核签字。工程质量评估报告主要内容包括：工程概况、承包单位基本情况、主要采取的施工方法、工程地基基础和主体结构的质量状况、施工中发生过的质量事故和主要质量问题及其原因分析和处理结果、对工程质量的综合评估意见。

（5）竣工验收。竣工验收完成后，由项目总监理工程师和建设单位代表共同签署《竣工移交证书》（附录B-B8），并由监理单位、建设单位盖章后，送承包单位一份。

1.1.4.4 质量问题和质量事故处理

（1）对可以通过返修或返工弥补的质量缺陷，应责成承包单位先编写质量问题调查报告，提出处理方案；监理工程师审核后（必要时经建设单位和设计单位认可），批复承包单位处理。处理结果应重新进行验收。

（2）对需要加固补强的质量问题，总监理工程师应签发《工程暂停令》（附录 B-B4），责成承包单位编写质量问题调查报告，由设计单位提出处理方案，并征得建设单位同意，批复承包单位处理。处理结果应重新进行验收。

（3）监理工程师应将完整的质量问题处理情况记录归档。

（4）施工中发生的质量事故，承包单位应按有关规定上报处理，总监理工程师应书面报告监理单位。

工程材料、构配件和设备质量控制的基本程序如图 1-3 所示。

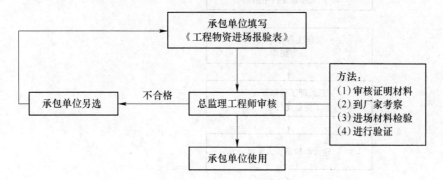

图 1-3　工程材料、构配件和设备质量控制基本程序

1.1.5　工程造价控制

1.1.5.1　工程造价控制基本程序

工程造价控制基本程序包括工程款支付基本程序和工程款竣工结算基本程序，分别如图 1-4 和图 1-5 所示。

1.1.5.2　工程造价控制的工作内容

（1）应严格执行工程计量和工程款支付的程序和时限要求，通过《工作联系单》（附录 C-C1）与建设单位、承包单位沟通信息，提出工程造价控制的建议。

（2）工程量计量，原则上每月计量一次，计量周期为上月 26 日至本月 25 日，承包单位应于每月 26 日前，根据工程实际进度及监理工程师签认的分项工程，上报月完成工程量。

（3）监理工程师对承包单位的申报进行核实，必要时应与承包单位协商，所计量的工程量应经总监理工程师同意，由监理工程师签认，对某些特定的分项、分部工程的计量方法则由项目监理部、建设单位和承包单位协商约定。

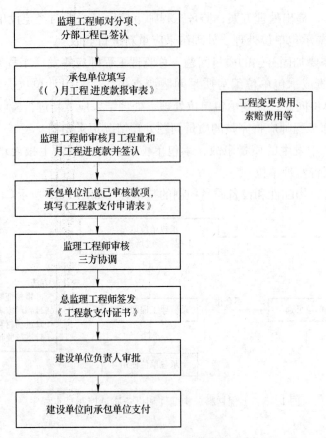

图1-4 工程款支付基本程序

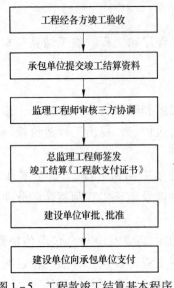

图1-5 工程款竣工结算基本程序

(4) 工程款支付分为"工程预付款"和"支付工程款"两个部分。承包单位填写《工程款支付申请表》，项目总监理工程师审核是否符合建设工程施工合同的约定，并及时签发工程预付款的《工程款支付证书》（附录B-B7），监理工程师应按合同的约定，及时抵扣工程预付款。监理工程师应要求承包单位根据已经计量确认的当月完成工程量，按建设工程施工合同的约定计算月工程进度款，并填写《( )月工程进度款报审表》（附录A-A11）。监理工程师审核签认后，应在监理月报中向建设单位报告，要求承包单位根据当期已发生且经审核签署的《( )月工程进度款报审表》《工程变更费用报审表》（附录A-A12）和

《费用索赔申请表》(附录A-A13)等计算当期工程款,并填写《工程款支付申请表》(附录A-A14)。监理工程师应依据建设工程施工合同及北京市有关规定,进行定额审核,确认应支付的工程款额度,由项目总监理工程师签发《工程款支付证书》,报建设单位。

(5) 工程竣工,经建设单位组织有关各方验收合格后,承包单位应在规定的时间内向项目监理部提交竣工结算资料,监理工程师应及时进行审核,并与承包单位、建设单位协商和协调,提出审核意见。总监理工程师根据各方协商的结论,签发竣工结算《工程款支付证书》,建设单位收到总监理工程师签发的结算支付证书后,应及时按合同约定与承包单位办理竣工结算有关事项。

1.1.6 施工合同其他事项管理

施工合同其他事项管理的内容包括:工程变更管理、工程暂停及复工管理、工程延期管理、费用索赔管理、合同争议协调和违约处理等。

1.1.6.1 工程变更管理

工程变更管理的程序如图1-6所示。

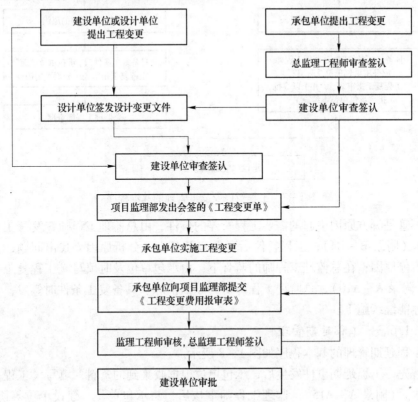

图1-6 工程变更管理的程序

工程变更要求可以由建设单位、设计单位、承包单位等提出，不管是哪方提出，都应填写《工程变更单》（附录C-C2），根据程序要求，变更实施后，承包单位应向项目监理部提出《工程变更费用报审表》（附录A-A12），以便结算使用。

1.1.6.2 工程暂停及复工管理

工程暂停及复工管理的基本程序如图1-7所示。

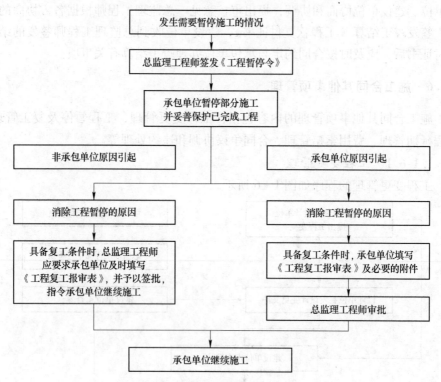

图1-7 工程暂停及复工管理的基本程序

不管是哪方原因引起的需要工程暂停的事件，由总监理工程师签发《工程暂停令》（附录B-B4）；工程暂停后联合有关各方，分析原因，找出问题，消除工程暂停原因；在总监理工程师的监督下，由承包单位及时填写《工程复工报审表》（附录A-A10），总监理工程师认为原因消除、具备复工条件时签发，承包单位才能继续施工。

1.1.6.3 工程延期管理

工程延期管理的基本程序如图1-8所示。

首先，工程延期事件发生后，承包单位根据收集到的资料，填写《工程延期申请表》（附录A-A15）；监理工程师审核后，与承包单位、建设单位等协商，回复同意或不同意意见；如同意，签发《工程延期审批表》（附录B-B5）。

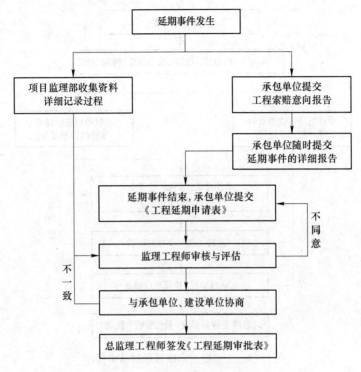

图1-8 工程延期管理的基本程序

1.1.6.4 工程费用索赔管理

工程费用索赔管理的基本程序如图1-9所示。

工程费用索赔事件发生后，首先，承包单位根据收集到的资料，填写《费用索赔申请表》（附录A-A13）；监理工程师审核后，与承包单位、建设单位等协商，回复同意或不同意意见；如同意，签发《费用索赔审批表》（附录B-B6）。

1.1.6.5 合同争议调解

合同争议调解的基本程序如图1-10所示。

合同争议事件发生后，争议一方或双方向项目监理部提出调解申请；接到调解申请后，总监理工程师根据调查情况，提出调解意见，并通过《工作联系单》（附录C-C1）通知双方进行调解；双方同意，调解成功，否则不成功，通过其他途径解决（不在管理范围）。

1.1.6.6 违约处理

违约处理的程序如图1-11所示。

首先，确认违约事件，违约事件发生后，相关方向监理部提出申诉报告；项目监理部调查、分析、取证，并通过《工作联系单》（附录C-C1）通知另一方，协商处理。

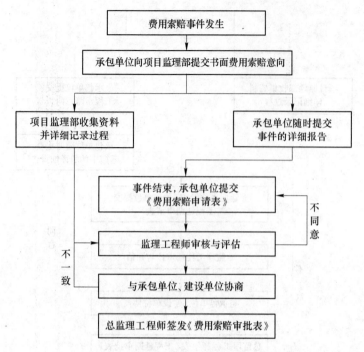

图 1-9 工程费用索赔管理的基本程序

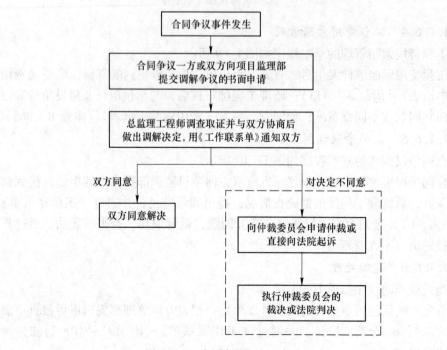

图 1-10 合同争议调解的基本程序

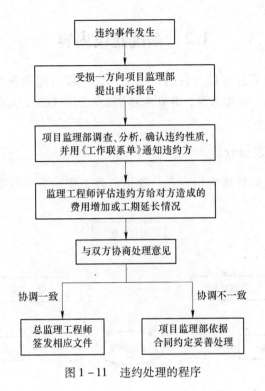

图 1-11 违约处理的程序

1.1.7 其他监理工作

其他监理工作包括工地会议、工程保修期监理、监理月报、监理工作总结、监理资料管理与归档和监理考核等。

工地会议分为"第一次工地会议"、"监理例会"和"专题工地会议"。第一次工地会议要求工程相关各方参加，协商监理有关事宜；监理例会由监理总工程师或监理工程师主持，定期（每周一次）召开；专题工地会议根据需要召开。

工程保修期要求监理做好定期回访记录。

监理月报的格式和内容参照附录 D，要求每月自动生成。

监理工作结束后，项目监理部向建设单位提交项目监理工作总结，内容包括：工程概况；监理组织机构、监理人员和投入的监理设备；监理合同履行情况；监理工作成效；施工过程中出现的问题及处理情况和建议；工程照片等。

监理资料管理与归档要求收集齐全，台账及时，随时呈报。

监理单位对项目监理部的考核要求按月进行。对考核中发现的问题应及时填写《不合格项处置记录》，要求项目监理部进行纠正，并制定相应有效的预防措施。根据问题数量和情节轻重应对当事人进行教育、批评、通报，直至撤换不称职的监理人员。

1.2 系统建设内容

建设工程监理信息建设开发的目标是建设工程监理业务管理信息化、智能化,实现异地办公、数据共享,并在此基础上,积累数据,运用数学方法,辅助决策。

1.2.1 监理业务逻辑分析

监理业务以工程管理为出发点,收集监理工程的有关数据。总体逻辑如图1-12所示。

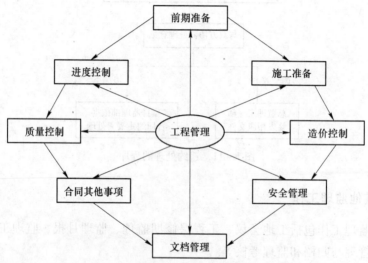

图1-12 建设工程监理业务的总体逻辑

1.2.1.1 工程管理业务逻辑

工程管理业务内容包括工程信息注册登记、工程状态调整、工程分项、信息修改和确认发布等,管理过程如图1-13所示。

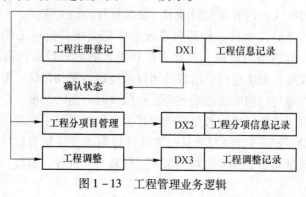

图1-13 工程管理业务逻辑

图 1-13 中，各个业务或操作是通过"数据存储"对象联结，这样有助于说明业务逻辑与数据存储之间的关系，同时说明业务所使用的数据存储对象，下文同，不再单独说明。

1.2.1.2 前期准备

施工监理的前期准备工作主要包括建立监理机构、选择相应工作人员、职责分工、准备必要的监理设备与设施、审阅施工图纸、编制工程项目监理规划和监理实施细则等，其业务逻辑如图 1-14 所示。

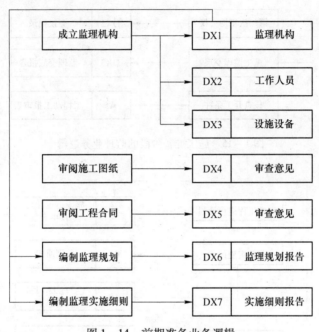

图 1-14 前期准备业务逻辑

1.2.1.3 施工准备

施工准备阶段的监理业务主要包括参与设计交底、审核施工方案、测量放线检查、组织工地会议、施工监理交底和核查开工条件等，其业务逻辑如图 1-15 所示。

1.2.1.4 进度控制

工程进度控制的主要业务包括审批进度计划、监督实施和计划调整等，其业务逻辑如图 1-16 所示。

1.2.1.5 质量控制

工程质量控制分为事前控制、事中控制、竣工验收和事故处理。事前控制包括核查质量管理体系、审查分包单位资质、查验测量放线情况、签认材料报验、签认建筑构配件及设备报验、检查进场施工设备、审查分部工程施工方案等；事

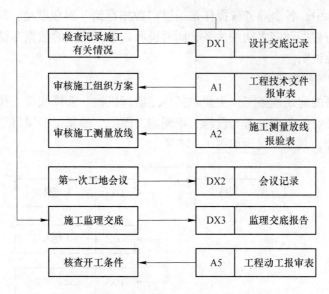

图1-15 施工准备阶段的监理业务逻辑

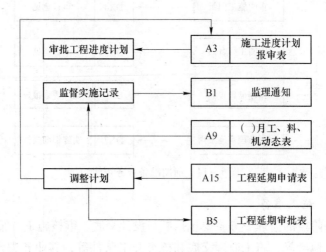

图1-16 工程进度控制业务逻辑

中控制包括现场巡视记录、核查工程预检、验收隐蔽工程、分项工程验收、分部工程验收等；工程竣工验收包括检查工程质量并记录、组织竣工验收、组织竣工移交等；事故处理包括日常巡视记录、事故处理等。工程质量控制业务逻辑如图1-17所示。

1.2.1.6 工程造价控制

工程造价控制分为工程计算、工程款支付、工程款竣工结算等业务，其业务逻辑如图1-18所示。

1.2 系统建设内容

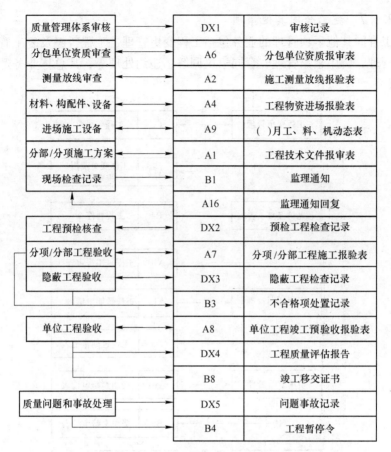

图 1-17 工程质量控制业务逻辑

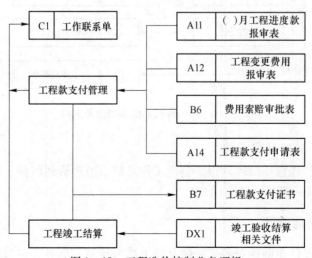

图 1-18 工程造价控制业务逻辑

1.2.1.7 施工合同其他事项

施工合同其他事项管理的业务包括工程变更管理、工程暂停及复工管理、工程延期管理、费用索赔管理、合同争议调解、违约处理等，其总体业务逻辑如图1-19所示。

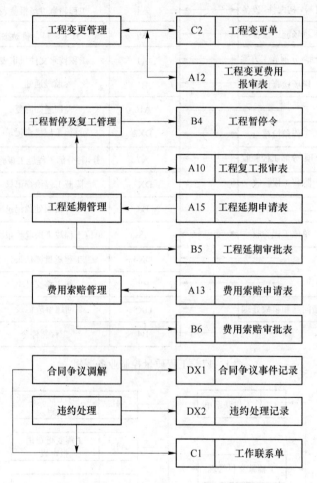

图1-19 施工合同其他事项业务逻辑

1.2.1.8 监理档案管理

监理档案管理包括监理月报、监理工作总结、监理资料归档、监理工作考核等业务，如图1-20所示。

1.2.2 系统功能设计

系统功能设计在前述业务逻辑的基础上进行抽象，另外，还需综合考虑辅助功能对系统运行的支持作用。系统总体功能如图1-21所示。

1.2 系统建设内容

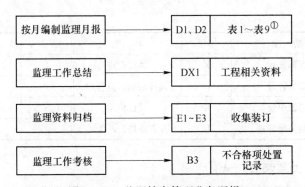

图1-20 监理档案管理业务逻辑

①各表的格式和内容请参考北京市地方性标准《建设工程监理规程》（DBJ 01-41—2002），后续章节同理

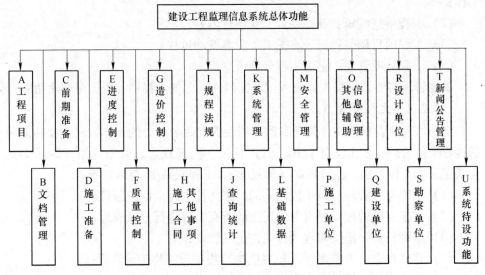

图1-21 建设工程监理信息系统总体功能

在此，增加"工程管理"、"文档管理"、"查询统计"、"基础数据"、"其他辅助信息管理"等项目，对系统总体功能进行补充和完善。另外，为了之后进行系统设计，总体功能分别以一位大写英文字母作为其代号，并按此方法进行扩展，其对照表列于表1-1中，供参考。

表1-1 功能字母对照表

字母	功能	字母	功能	字母	功能	字母	功能
A	工程项目	B	文档管理	C	前期准备		
D	施工准备	E	进度控制	F	质量控制		
G	造价控制	H	施工合同其他事项	I	规程法规		

续表1-1

字母	功能	字母	功能	字母	功能
J	查询统计	K	系统管理	L	基础数据
M	安全管理	O	其他辅助信息管理	P	施工单位
Q	建设单位	R	设计单位	S	勘察单位
T	新闻公告管理	U	系统待设功能		

1.2.2.1 工程项目（A）

工程管理主要负责工程信息建档、工程增加、工程项目调整、工程分项管理等任务，并设置专职岗位，进行操作管理，该岗位主要工作内容包括：

（1）增加新的工程：通过提供工程编号、工程名称等部分信息，增加新的工程；

（2）工程信息编辑：补充完善工程信息内容；

（3）工程项目调整：记录有关工程内容变动过程；

（4）工程分项管理：记录工程分项信息。

根据管理需要，确保工作流程记录完整性，此处不提供工程记录删除功能。

1.2.2.2 文档管理（B）

文档管理负责日常文件管理和监理档案管理两个部分的工作，其主要功能包括接收文件管理、发出文件管理、监理日记、监理日志、监理月报（附录D）、会议记录、文件类别、监理档案（附录E）、工程联系单（CA）等。

（1）接收文件管理：对单位以外的外来文件进行全面记录管理；

（2）发出文件管理：对本单位发出的所有文件进行记录管理；

（3）监理日记：提供监理员日工作记录管理；

（4）监理日志：为监理部（监理机构）提供日工作记录管理；

（5）会议记录：提供所有有关监理会议记录管理；

（6）文件类别：文件往来管理辅助功能；

（7）监理月报：对监理报告（附录D）进行管理；

（8）监理档案：对监理档案（附录E）进行管理；

（9）监理发文：对有关监理工作所发出的工作联系记录进行管理。

1.2.2.3 前期准备（C）

前期准备功能包括监理机构管理、监理人员管理、设施设备管理、参与单位管理、合同查阅记录、图纸查阅记录、监理规划和监理细则管理等，工程施工图纸信息记录功能也包括其中。

（1）监理机构管理：提供工程监理机构的设置记录管理；

（2）监理人员管理：设置工程监理工作相关人员的信息记录；

（3）设施设备管理：对工程监理机构所使用的设施和设备进行记录管理；

（4）参与单位管理：对与工程建设、施工等有关的参与单位进行记录管理；

（5）合同查阅管理：记录工程建设合同和工程施工合同的审查过程及发现的问题；

（6）图纸查阅管理：记录对工程施工图纸的审查情况；

（7）监理规划：记录所制订的工程监理规划及方案；

（8）监理细则：记录所制订的工程监理详细实施细则内容；

（9）工程图纸管理：记录工程施工有关的图纸信息内容。

1.2.2.4 施工准备（D）

施工准备功能包括参与设计交底管理、审查相关技术文件报审表、审查测量放线报验表、第一次工地会议管理、施工监理交底管理、审批工程动工报审表等工作。

（1）参与设计交底：记录管理对施工单位开工之前所进行的各项检查情况；

（2）技术文件报审-AA：记录管理对施工单位报审的《工程技术文件报审表》（附录A-A1）的审查情况；

（3）测量放线报验-AB：记录管理对施工单位报审的《施工测量放线报验表》（附录A-A2）的审查情况；

（4）首次会议纪要：记录管理第一次工地会议情况，并整理成报表；

（5）施工监理交底：记录管理所制订的监理工作实施方案，并准备报告；

（6）工程动工报审-AE：记录管理对施工单位报审的《工程动工报审表》（附录A-A5）的审查情况。

1.2.2.5 进度控制（E）

工程进度控制功能包括进度计划审核、工料机动态审核、延期申请审核、签发监理通知、签发延期审批等任务。

（1）进度计划审核-AC：对施工单位定期上报的《施工进度计划报审表》（附录A-A3）进行审核，并签署意见；

（2）工料机动态审核-AI：对施工单位定期上报的《（ ）月工、料、机动态表》（附录A-A9）内容进行审核，并签署意见；

（3）延期申请审核-AO：对施工单位上报的《工程延期申请表》（附录A-A15）进行审核，并签署意见；

（4）监理通知-BA：发现进度计划执行存在严重问题时，及时向施工单位发出《监理通知》（附录B-B1），提出意见；

（5）延期审批-BE：签发并管理《工程延期审批表》（附录B-B5）。

1.2.2.6 质量控制（F）

工程质量控制功能包括管理体系审核、分包单位审核、测量放线审核、进场物资检查、施工设备检查、分项/分部施工方案、监理通知、监理通知回复、工程预检审核、分项/分部验收、隐蔽工程验收、不合格项处理、单位工程验收、质量评估、工程竣工移交、质量问题和事故处理、工程暂停令、工程图片管理等任务。

（1）管理体系审核：对承包单位有关资质和质量管理体系进行审核记录；

（2）分包单位审核－AF：对分包单位的有关资质通过《分包单位资质报审表》（附录A－A6）进行审核；

（3）测量放线审核－AB：审核管理《施工测量放线报验表》（附录A－A2）及其内容；

（4）进场物资检查－AD：审核管理《工程物资进场报验表》（附录A－A4）及其内容；

（5）施工设备检查－AI：审核管理《（ ）月工、料、机动态表》（附录A－A9）及其内容；

（6）分项/分部施工方案－AA：审核管理《工程技术文件报审表》（附录A－A1）及其内容；

（7）监理通知－BA：对现场检查发现的问题发出《监理通知》（附录B－B1），要求整改；

（8）监理通知回复－AP：管理《监理通知回复单》（附录A－A16）及其内容；

（9）工程预检审核：管理审核施工单位上报的工程预检报告；

（10）分项/分部验收－AG：审核管理《分项/分部工程施工报验表》（附录A－A7）及其内容；

（11）隐蔽工程验收：记录管理对隐蔽工程验收的有关数据；

（12）不合格项处理－BC：管理《不合格项处置记录》（附录B－B3）及其内容；

（13）单位工程验收－AH：审核管理《单位工程竣工预验收报验表》（附录A－A8）及其内容；

（14）质量评估：根据要求，管理工程质量报告；

（15）竣工移交－BH：管理《竣工移交证书》（附录B－B8）及其内容；

（16）问题事故：记录对质量问题和工程事故的处理记录；

（17）工程暂停令－BD：管理《工程暂停令》（附录B－B4）及其内容；

（18）工程图片管理：记录工程施工的有关图片内容。

1.2.2.7 造价控制（G）

工程造价控制功能包括工程款支付管理和竣工结算管理两个部分，具体分为工作联系、审核月工程进度款报审表、审核工程变更费用报审表、审核费用索赔表、审核工程款支付申请表、审核工程竣工结算申请、签发工程款支付证书等任务。

（1）工作联系单-CA：管理签发的《工作联系单》（附录C-C1），保存沟通资料；

（2）月进度款报审-AK：管理施工单位上报的《（ ）月工程进度款报审表》（附录A-A11）；

（3）变更费用报审-AL：管理施工单位上报的《工程变更费用报审表》（附录A-A12）；

（4）费用索赔申请-AM：管理施工单位上报的《费用索赔申请表》（附录A-A13）；

（5）工程款支付申请-AN：管理施工单位上报的《工程款支付申请表》（附录A-A14）；

（6）竣工结算申请：管理对施工单位报送的有关工程竣工结算材料并记录；

（7）工程款支付证书-BG：管理签发的《工程款支付证书》（附录B-B7）。

1.2.2.8 施工合同其他事项（H）

"施工合同其他事项"是指施工合同其他事项的管理，其内容包括工程变更管理、工程暂停及复工管理、工程延期管理、费用索赔管理、合同争议调解、违约处理，具体分为审核工程变更单、审核工程变更费用报审表、签发工程暂停令、管理工程复工报审、审核工程延期申请、签发工程延期审批表、审核费用申请表、签发费用索赔审批表、记录合同争议事件、记录违约处理内容、签发工作联系单等。

（1）工程变更单-CB：管理来自各方的《工程变更单》（附录C-C2）内容；

（2）变更费用报审-AL：审核管理施工单位报送的《工程变更费用报审表》（附录A-A12）；

（3）工程暂停令-BD：签发《工程暂停令》（附录B-B4）；

（4）工程复工报审-AJ：管理审核《工程复工报审表》（附录A-A10）；

（5）工程延期申请-AO：管理审核《工程延期申请表》（附录A-A15）；

（6）工程延期审批-BE：管理签发《工程延期审批表》（附录B-B5）；

（7）费用索赔申请-AM：管理审核《费用索赔申请表》（附录A-A13）；

（8）费用索赔审批-BF：管理签发《费用索赔审批表》（附录B-B6）；

（9）合同争议事件：记录管理合同争议事件及调解内容；

（10）违约处理记录：记录管理违约处理内容；

（11）工作联系单-CA：管理签发《工作联系单》（附录C-C1）。

1.2.2.9 规程法规（I）

规程法规功能主要完成有关法律、法规、政策、规定等有关文件的管理，具体分为文件管理、文件上传、文件下载、文件查询等功能。

(1) 文件管理：收集整理相关文件并记录；
(2) 文件上传：上传相关文件内容；
(3) 文件下载：提供文件下载功能；
(4) 文件查询：提供文件查询功能。

1.2.2.10 查询统计（J）

查询统计功能是为了满足数据管理和决策管理需要和监理档案管理需要，特别增加的监理业务以外的功能，目前设置内容为工程项目统计、管理表格统计、信息回复统计、监理工程统计、年份工程量统计、工程款审批及支付汇总统计，并且可以在现有数据文件的基础上，根据管理和用户需要，适当扩展和增补。

(1) 工程项目统计：以年月日为时间轴对工程数据进行统计，包括年工程量、月工程量、日工程量、同期比较等；
(2) 管理表格统计：统计工作报表的数量，例如监理日记、监理日志等；
(3) 信息回复统计：统计施工单位对监理通知的回复数量；
(4) 监理工作统计 – DI：根据时间轴，对附录 D – D9 中的各项内容要求进行统计；
(5) 年工程量分类统计：以工程的不同属性为类别进行工程量统计，并用图表方式展示；
(6) 工程款审批及支付汇总统计 – DH：完成工程款审批及支付汇总表（附录 D – D8）所要求项目的统计工作。

1.2.2.11 系统管理（K）

系统管理功能包括系统角色管理、系统功能管理、角色功能管理、修改用户口令、操作日志管理等内容。

(1) 系统角色：管理系统角色，并通过角色确定人员操作权限；
(2) 系统功能：管理系统功能，提供操作导航；
(3) 角色功能：管理角色的功能，提供操作的分类实现；
(4) 重置密码：对用户的密码重新设置；
(5) 操作日志：管理用户操作记录。

1.2.2.12 基础数据（L）

基础数据功能主要完成常规的共用数据的管理任务，包括往来单位管理、用户职员管理、单位部门管理、设施设备管理、工程状态管理、工程类别管理、工程性质管理、工程级别管理、单位类型管理、报表目录管理、合同类别管理、合同管理等。

（1）往来单位：管理与工程建设施工监理业务有关的单位信息；
（2）用户职员：管理系统用户和单位职工信息（职工视为用户进行管理）；
（3）单位部门：管理本单位部门信息；
（4）设施设备：设施和设备信息管理；
（5）工程状态：管理工程状态信息；
（6）工程类别：管理工程类别信息；
（7）工程性质：管理工程性质信息；
（8）工程级别：管理工程级别信息；
（9）单位类型：管理单位类型信息；
（10）工作任务：记录管理监理工作任务内容，供查阅；
（11）报表目录：管理监理报表目录，提供处理入口；
（12）文章类别：管理有关新闻公告、用户建议等的类别信息；
（13）新闻公告：管理新闻公告内容；
（14）合同类别：管理合同类别信息；
（15）合同管理：管理有关合同信息，供查阅。

1.2.2.13 安全管理（M）

安全管理功能包括施工单位资质审核、施工方案审核、安全管理细则交底、工作联系单管理、工程暂停令管理、安全问题事故处理等任务。

（1）施工单位资质：记录管理对施工单位有关资质的审查情况；
（2）施工方案审核：记录管理对施工单位所有的施工方案的审查情况；
（3）安全管理交底：编制记录管理施工安全细则；
（4）工作联系单 – CA：管理签发《工作联系单》（附录 C – C1）；
（5）工程暂停令 – BD：签发《工程暂停令》（附录 B – B4）；
（6）安全问题事故处理：记录管理有关安全问题和事故处理的情况。

以上列出了系统应有的功能，在具体开发过程中，有可能会修改或合并，也有可能增加，因此实际完成后的系统会有变化。

1.2.3 监理组织机构

监理组织机构包括监理单位组织机构和临时项目监理机构。

1.2.3.1 监理单位组织机构

监理单位组织机构一般由经理办公室、人力资源部、技术管理部、财务结算部、业务联系部、招标投标部、档案管理部等部门组成，如图 1 – 22 所示。

从监理单位的业务关系出发，可以将所有业务重新组合并分为综合事务管理、人力资源管理、工程技术管理、监理业务管理和财务审计管理。因此，上述组织机构可简化为综合事务部、人力资源部、工程技术部、监理业务部、财务审

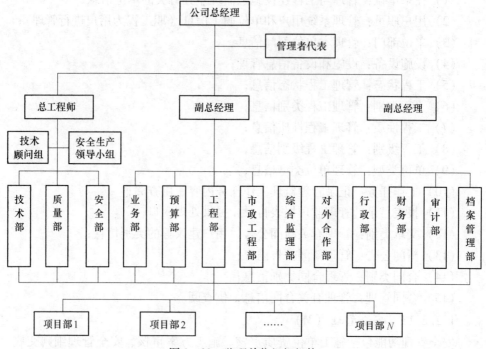

图1-22 监理单位组织机构

计部,并增加信息技术部可负责档案管理工作,增加安全监督委员会代替各种监督和决策参议职能,如图1-23所示。

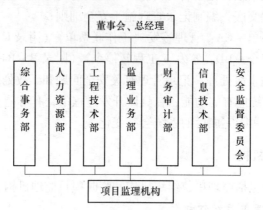

图1-23 合并后的监理单位组织机构

1.2.3.2 项目监理机构

由监理单位根据监理合同要求、派驻工程项目施工工地、负责履行委托监理合同的组织机构,一般称为"项目监理部"、"监理公司分部"、"监理公司分处"等。项目监理机构是由所在的监理单位组织并派驻到监理工程现场。项目监理机构是一次性的，在完成委托监理合同约定的监理工作后即行解体，其组织机构如图 1-24 所示。

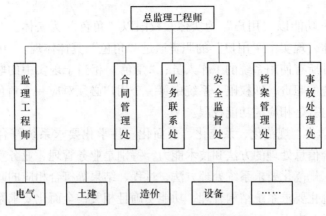

图 1-24　项目监理机构组织结构

项目监理机构人员的配备构成，随着项目的监理阶段不同而异，总体结构组成包括：总监理工程师、总监理工程师代表、专业监理工程师和监理员、专职或兼职的安全监督员、合同管理员、资料管理员及必要的辅助工作人员。

一名总监理工程师只宜担任一项委托监理合同项目的总监理工程工作，如果需要同时担任多个委托监理合同的项目总监理工程师时，需要经过建设单位同意，并且最多不得超过 3 个项目。

监理工程师可根据工程项目的性质由不同专业和资质资格的工程师组成，可在各个项目监理机构同时开展工作。

1.2.4　系统设计思想

建设工程监理业务是一个复杂的系统工程，其特点是动态性强、数据量大、交叉回合多、涉及范围广、要求处理速度快。建设工程监理信息系统必须充分考虑业务特点和数据处理要求，从信息处理的关键处入手，使设计和开发工作更加系统化、科学化、全面化。

（1）以"工程项目"为中心，建立新型业务系统。建设工程监理业务是一个复杂的网络工程，根据业务流程和数据处理流程不难发现，"工程项目"是整体业务体系和业务管理的"中心"对象，以"工程项目"为中心，建立业务处理功能，是监理信息系统建设的基础。同时，数据处理的组织和管理是数据库建设的出发点，如图 1-12 所示。

（2）功能合理分解，减少个体工作量。建设工程监理业务涉及建设单位、施工单位、设计单位、监理单位等，同时项目监理机构要求人员动态管理功能强，办公地点流动性强，业务处理分散且时性要求较高。因此，系统功能设计时，需要相应单位和人员完成的业务，在系统中也需要分别完成，并用相应的安全逻辑进行合理控制，例如有关工程监理所需要的各种报审材料，尽量由施工单位完成。

（3）业务功能以"用户"为实体，权限以"角色"为实体。"用户"是功能实现的主体，现实中"用户"的职责通过"角色"具体体现。"用户"包括单位内和单位外所有使用系统的工作人员；"角色"相当于现实中的职责，不同的角色具有相应的职责，即权限；系统中每一个用户必须对应一个角色，以体现相应的职责，也就是相应的功能权限。

（4）科学化、规范化、系统化、全面化。科学化要求系统符合信息化技术的要求，符合信息处理的方法和技术能力，特别是业务管理、业务重组符合现有业务的要求；规范化要求系统处理方法、过程、结果等符合相应的政策法规和管理规范；系统化要求系统结构合理、功能明确且相关；全面化要求系统处理覆盖监理业务的各个方面。

1.2.5　系统设计要求

系统设计的指导思想是利用先进的信息技术（计算机数据处理技术、网络数据传输技术、数据库技术），根据工程建设施工质量管理系统的发展现状，结合具体的监理业务管理体系，建立建设工程监理信息系统，并以此为起点，建设工程监理业务管理信息化、高效化，并向智能化和智慧化方向发展。系统设计的具体要求如下：

（1）**建设目标明确**。系统建设的目标是：利用信息技术建立符合监理业务管理要求并高效的监理信息系统，以实现监理业务管理的网络化和规范化。

根据建设工程监理相关政策法规和规程要求，基于监理企业监理业务基础，以项目管理理论与方法为指导，建立建设工程监理信息系统，将监理单位的监理管理工作和监理单位的有关信息集合于一体，并利用网络技术，实现异地管理，提高工程项目施工质量监理效率，为施工单位和监理单位等相关部门提供方便、准确的信息，从而实现工程施工每个阶段的质量控制高效、合理、规范。

（2）**系统设计的原则**。通常信息系统在设计上应满足以下几个方面的要求：

1）**实用性**：系统开发必须采用成熟的技术，认真细致地做好功能和数据的分析，并充分利用现代信息技术，合理地融合于业务逻辑，力求向用户提供符合管理规则要求和现行管理方法要求的功能完整的信息处理系统。

2）**开放性**：系统要实现网络互联、资源共享、多用户访问和二次开发。

3）**先进性**：运用成熟并且先进的技术、编程语言、开发工具、设计思想。

4）**安全性**：由于施工项目的规模大，导致施工单位的申报程序繁琐和监理单位对工程质量监理的项目多。因此，要求数据传输的可靠性和保密性。

5）可扩展性：目前，我国施工质量管理不断规范化、智能化，利用计算机进行的管理业务范围将不断扩大，因此，系统必须具备充分的扩展能力。

6）完整性：实现优化的网络设计，安全可靠的数据管理，高效的信息管理，友好的用户界面。

（3）系统运行模式。企业信息系统的运行模式可以是C/S、B/S或二者的混合结构。

C/S，即客户/服务器结构模式，是一种两层结构的系统，第一层是在客户机系统上结合了表示与业务逻辑，第二层是通过网络结合了数据库服务器，适合局域网络环境应用。

B/S，即浏览器/服务器结构模式，客户通过浏览器向WEB服务器发出请求，WEB服务器向数据库服务器发出数据需求，WEB服务器将结果返回浏览器，这是三层结构，适合于远程业务信息的实现，是系统建设的首选结构模式。

（4）开发工具和运行环境。系统建设开发工具选用Microsoft Visual Studio 2013，简称为"VS2013"；数据库管理系统选用Microsoft SQL Server 2012；辅助技术还有HTML5、CSS3、MVC、EF等；WEB服务系统选用Microsoft Internet Information Service – IIS；WEB运行平台是ASP. NET 4.5或更高。

1.3　管理对象分析

信息系统建设的核心是业务对象及其逻辑关系的分析与抽象，是信息系统的处理对象，是建设信息系统数据库和信息系统智能化的基础。通过建设工程监理业务过程，研究分析并从中抽取监理业务所涉及的对象及其关系、建立监理信息系统的"数据模型"，是建设工程监理信息建设开发的前提。本节内容以"监理工程"对象为中心，研究建设工程监理信息系统所涉及和需要管理的对象及其关系与性质。

1.3.1　监理工程对象及属性

"工程"是监理业务的核心管理对象。这里所指的工程，是监理单位根据合同所承接的监理工程，包括投标中标工程、指派工程、援助工程、自建工程、委托工程等。根据建设工程的特征和监理管理业务的需要，并适当兼顾监理业务有关统计分析工作的需要，以内容全面、结构合理为出发点，监理的"工程"对象属性分为基本属性、业务属性和相关属性。

（1）工程基本属性。包括工程名称、施工地点（工程地址）、监理开始日期（注册登记日期）、经手人员、项目经理、开工日期、竣工日期、功能说明、质量目标、合同价款、合同模式、规划、规划许可证号、建设、建设许可证号、开工许可、开工证号、投资性质、投资主体、建设面积、报建面积、监理费用、调整日期、调整人员、开始桩号、结束桩号、工程级别、工程性质、工程状态、工程类别、备注说明。

（2）工程附加属性。特别附加属性为工程编号，为检索、查询、存储等操作设置的工程附加属性，也是未来信息系统中工程的唯一识别属性，要求以年号和顺序号相结合的方法取值。

（3）相关人员属性。工程施工和监理过程中，会有不同类别的管理人员，同时根据需要，还会有增加、减少、职责变动的情况发生。从动态管理需求出发，需要单独设置相关人员属性。工程相关工作人员对象属性项目设置有：序号、识别号、姓名、性别、职务、职责、开始工作时间、结束工作时间等。

（4）工程调整。工程调整是对监理过程中工程变化的记录，是动态的，以独立对象方式管理，其相应的属性项目有：序号、调整内容、调整时间、经手人、负责人等。

（5）工程级别。工程级别反映工程的层次和规模，例如，国家重点建设工程、大中型公用事业工程、成片开发建设的住宅小区工程、利用外国政府或者国际组织贷款、援助资金的工程等，需要以单独对象方式管理。其属性项目有：级别编号、级别名称、相关说明等。

（6）工程性质。工程性质反映工程来源及内容范围，例如，新建项目、扩建项目、迁建项目、恢复项目、市政/公用工程项目、政府投资兴建和开发建设的办公楼、社会发展事业项目和住宅工程项目更新改造项目，包括挖潜工程、节能工程、安全工程、环境保护工程等，还有外资、中外合资、国外贷款、赠款、捐款建设的工程项目。其属性项目有：性质编号、性质名称、相关说明等。

（7）工程类别。工程类别反映工程的建筑形态，例如，工业建筑工程、民用建筑工程、构筑物工程、单独土石方工程、桩基础工程等。其属性项目有：类别编号、类别名称、相关说明等。

1.3.2 监理业务管理信息交换记录对象

监理业务管理信息交换记录对象由北京市地方性标准《建设工程监理规程》（DBJ 01-41—2002）中的附录 A~E 中的表格为依据。将每个表格视为一个管理对象，由于依附于工程，是工程对象的业务扩展对象，其共有的属性包括序号、工程编号、文件编号、文件日期等，对于每个对象的特有属性将在后续章节中加以详细说明。

1.3.3 系统服务对象

系统服务对象主要指人员（用户和工作人员）、角色（权限或级别）、系统功能等对象。

（1）人员。包括系统用户和监理单位工作人员，为了方便说明，统一称为"系统用户"对象，方便进行统一化管理。对象属性定义如下：

1）基本属性：用户标识、职员姓名、性别、出生日期、证件号码（身份证号）、职务、职称、专业、通信地址、家庭住址、邮政编码等；

2）工作属性：职责（角色）、工作单位、本单位部门、联系电话、入职时间、通信地址、离职时间等；

3）用户属性：登录次数、用户密码、QQ号码、微信账号、照片目录、用户照片、照片类型等。

（2）角色。角色是系统记录用户职责或权限的对象，通过角色，每个系统用户具备相应的职责或权限，进行不同的操作和信息获取范围。其属性有：角色编号、角色名称、备注说明等。

（3）系统功能。系统功能对象反映信息系统可能具备的能力，每种能力通过特定的功能加以表现，并以可操作的方式提供给系统用户使用。系统功能对象根据信息系统开发所使用工具或架构可能有不同属性组合。本系统定义的属性有：功能编号、功能名称（显示名称）、备注说明、方法名称、控制器名称、区域名称、方法参数值等。

（4）角色功能。从实体管理的角度出发，系统角色对象和系统功能对象存在关联关系，即角色的职责是通过相应的功能体现；反之，系统功能通过系统角色加以实现。一个角色实例可以具备多个系统功能实例；一个功能实例可以被多个角色实例拥有，因而两者是多对多关系。角色功能也可以将二者之间的关系转化为一对多或多对一关系。角色功能关系属性有：角色编号、功能编号、记录标识号等。

1.3.4 辅助数据对象

辅助数据对象也是系统服务对象，主要任务是反映和记录系统中所使用的基础数据和常用不变的数据，根据实际需要可以增加必要的数据对象，下面列出几个常用的辅助数据对象。

（1）设施设备。设备对象是指与监理业务相关并且可用的设施设备实体，其属性有：设备编号、设备名称、功能说明、规格型号、库存数量、购置日期、设备单价、计量单位、备注说明等。

（2）往来单位。往来单位对象是指与建设工程、监理业务相关的实体，其属性有：单位编号、单位名称、单位地址、单位类型、联系人员、负责人、联系电话、备注说明等。

（3）往来单位类型。往来单位类型是动态反映往来单位特征和性质的延伸对象，服务于往来单位对象，其属性有：单位类别编号、单位类别名称、备注说明等。

另外，还有许多辅助数据对象，例如之前已介绍过的工程性质、系统用户和人员、工程级别等，根据系统需要，可随时增加所需要的辅助数据对象。

本章小结

本章内容从监理业务分析入手，分析了监理业务及其业务流程逻辑、系统管理可能涉及的对象模型，并提出了系统建设的具体内容和要求。

附　录[1]

附录A　承包单位用表

A1　工程技术文件报审表
A2　施工测量放线报验表
A3　施工进度计划报审表
A4　工程物资进场报验表
A5　工程动工报审表
A6　分包单位资质报审表
A7　分项/分部工程施工报验表
A8　单位工程竣工预验收报验表
A9　（　）月工、料、机动态表
A10　工程复工报审表
A11　（　）月工程进度款报审表
A12　工程变更费用报审表
A13　费用索赔申请表
A14　工程款支付申请表
A15　工程延期申请表
A16　监理通知回复单

附录B　监理单位用表

B1　监理通知

---

[1] 有关各表的具体格式和填写内容请参考北京市地方性标准《建设工程监理规程》（DBJ 01-41—2002）。

B2　监理抽检记录
　　B3　不合格项处置记录
　　B4　工程暂停令
　　B5　工程延期审批表
　　B6　费用索赔审批表
　　B7　工程款支付证书
　　B8　竣工移交证书

附录 C　各方通用表
　　C1　工作联系单
　　C2　工程变更单

附录 D　监理月报格式
　　D1　监理月报封面
　　D2　监理月报目录

附录 E　工程监理档案
　　E1　工程监理档案封面
　　E2　工程监理档案移交目录
　　E3　工程监理档案审核备考表

2 建立工程监理信息系统项目

Visual Studio 是目前流行的基于 Windows 平台的应用程序集成开发工具，目前版本为 Visual Studio 2013。"建设工程监理信息系统"项目是使用 Visual Studio 2013 工具建设开发而成的基于 .NET Framework 4.5 及以上版本运行架构的 Web 项目。本章内容如下：

2.1　Visual Studio 2013 简要概述
2.2　建立监理信息系统项目
2.3　MVC 目录架构
2.4　Web.config 文件

2.1　Visual Studio 2013 简要概述

Visual Studio 2013 是 Microsoft 公司在 Visual Studio 2012 推出不到一年时间里的更新版本（下文中简称 VS2013）。

2.1.1　VS2013 的主要新功能

VS2013 的主要新功能表现在以下几个方面：

（1）支持 Windows 8.1 App 开发。VS2013 提供的工具非常适合利用下一代 Windows 平台（Windows 8.1）生成新式应用程序，同时在所有 Microsoft 平台上支持设备和服务。支持在 Windows 8.1 中开发 Windows App 应用程序，具体表现在：对工具、控件和模板进行了许多更新；对于 XAML 应用程序支持新近提出的编码 UI 测试；用于 XAML 和 HTML 应用程序的 UI 响应能力分析器和能耗探查器；增强了用于 HTML 应用程序的内存探查工具以及改进了与 Windows 应用商店的集成。

（2）敏捷项目管理（Agile Portfolio Management）。提供敏捷项目组合管理，提高团队协作，TFS2012 已经引入了敏捷项目管理功能，在 TFS2013 中该功能将得到进一步改进与完善（比如 backlog 与 sprint）。TFS 将更擅长处理流程分解，为不同层级的人员提供不同粒度的视图 backlog，同时支持多个 Scrum 团队分开管理各自的用例backlog，最后汇总到更高级的backlog。这意味着TFS将更重视企业敏捷，相信在新版本中还将提供更完善的敏捷支持。

在得到有效应用的情况下，ALM实践方法可以消除团队之间的壁垒，使企业能够克服挑战，更快速地提供高质量的软件。采用ALM的公司还可以减少浪费、缩短周期时间和提高业务灵活性，从而受益。

（3）版本控制。在近几个版本中VS一直在改进自身的版本控制功能，包括Team Explorer新增的Connect功能，可以帮助你同时关注多个团队项目。新的Team Explorer主页也更简洁、明确，在各任务间切换变得更加方便。同时，由于众多用户反馈，VS2013中将恢复更改挂起（Pending Changes）功能。如果你对VS、TFS有什么建议或者意见，也可以考虑向VS开发团队反馈。

（4）轻量代码注释（Lightweight Code Commenting）。其与VVS高级版中的代码审查功能类似，可以通过网络进行简单的注释。

（5）编程过程。新增代码信息指示，在编程过程中，VS2013增强了提示功能，能在编码的同时帮助监查错误，并通过多种指示器进行提示。此外，VS2013中还增加了内存诊断功能，对潜在的内存泄露问题进行提示。

（6）测试方面。在VS/TFS2012中测试功能已经有不少改进，VS/TFS2013更进一步完善了该功能，比如VS2012中引入的基于Web的测试环境得到了改进。

VS2013中还新增了测试用例管理功能，能够在不开启专业测试客户端的情况下测试计划进行全面管理，包括通过网络创建或修改测试计划、套件以及共享步骤。自2005版以来，VS已经拥有了负载测试功能，VS2013中的云负载测试大大简化了负载测试的流程。

（7）发布管理。近些年，产品的发布流程明显更加敏捷，因此很多开发者需要更快、更可靠并且可重复地自动部署功能。在刚刚结束的TechEd大会上，微软宣布与InCycle Software Inc达成协议，将会收购后者旗下的发布管理工具In-Release。因此，InRelease将会成为TFS原生发布解决方案。

（8）团队协作。顾名思义，TFS的核心任务之一就是改进软件开发团队内部的协作，TFS2013中将新增"Team Rooms"来进一步加强该特性，登记、构建、代码审查等一切操作都将会被记录下来，并且支持代码评论功能。

（9）整合微软System Center IT管理平台。除此之外，VS2013还有团队工作室、身份识别、.NET内存转储分析仪、Git支持等特性，可以看出这次将团队合作作为了一个重要的部分，结合Windows Azure云平台进行同步协作。

2.1.2 VS2013开发环境

VS2013开发环境是一个集成的IDE环境，和传统IDE操作功能相比，其最大的特色是集成了 NuGet 程序包管理器，开发者在统一的 IDE 平台中即可完成第三方扩展工具的查询、增加、扩展等管理。VS2013 开发环境如图 2-1 所示。

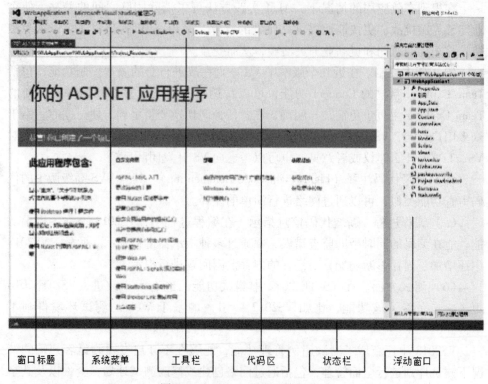

图 2-1　VS2013 开发环境概览图

下面对常用功能区加以说明。

（1）系统菜单。系统菜单是传统的下拉方式，由文件、编辑、视图、项目、生成、调试、团队、工具、测试、体系结构、分析、窗口和帮助等主项目组成。

1）新建（F）：建立新的 VS2013 工程项目，根据向导，项目类别可选择现有安装模板或从网络下载模板两种方式；既可以是 Web 项目，也可以是 Windows 项目。

2）打开（O）：打开已有项目进行设计与开发，也可以从"最近使用的项目和解决方案"功能项中选择最近常用的项目（默认列出最近常用的 4 个项目）。

3）编辑（E）：常用的文本编辑操作。

4）视图：包括解决方案资源管理器、服务器资源管理器等各种悬浮窗口的显示和隐藏。

5）项目：常用子项目是项目属性，打开当前项目有关的属性和参数定义窗口，例如，调试用 IIS 服务器的参数定义、项目版本参数定义等。

6) 生成：常用子项目是生成、清理、发布功能。生成是运行前的"编译"，解决新修改或新增加内容的及时更正作用；清理是对现有信息提示和运行状态恢复；发布是本地项目内容发送至远程服务器并更新，可整体或单项发布。

7) 调试：启动调试和停止调试，项目运行过程中的意外故障，可通过"停止调试"功能终止，返回设计编辑状态。

8) 工具：常用的是 NuGet 程序包管理器，可以更新系统组件和第三方控件；还有选项，打开系统有关的环境参数定义窗口。

其他菜单项目及功能请读者参考有关资料。

(2) 工具栏。工具栏上的项目是菜单栏中部分功能项目的快捷方式，数量少于菜单栏，可以通过自定义方式增加所需要的快捷功能项目。工具栏的内容会根据当前所编辑的项目内容不同而变化，常用的是新建、打开、剪切、复制、粘贴、调试、存储、撤销等功能操作。

(3) 解决方案资源管理器。解决方案资源管理器窗口位于界面的右边，以目录树的方式显示当前解决方案所包含的项目内容，是整个项目内容集中显示的选择区，其形式如图 2-2 所示。

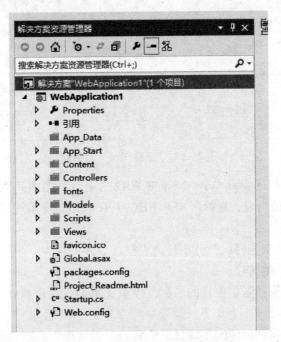

图 2-2　解决方案资源管理器

2.1.3　VS2013 新建项目

VS2013 通过模板建立新项目时，首先分为最近、已安装和联机三种获取方式；根据语言不同，分为 Visual C#、Visual Basic、Visual F#、JavaScript 项目等；在此基础上，再选择 Web 项目或 Windows 项目。现以"已安装—Visual C#—Web"项目为例，说明 VS2013 新建项目的方法和过程。

2.1.3.1 新建项目

文件菜单选择"新建项目"，打开"新建项目"对话框，如图 2-3 所示。

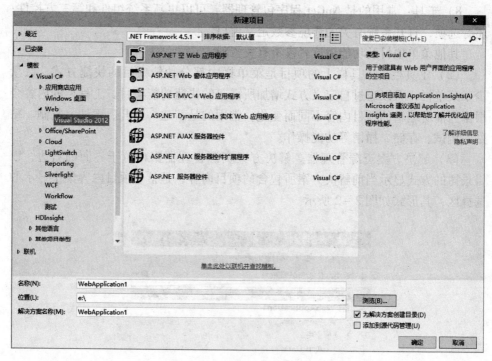

图 2-3 "新建项目"对话框

选择"Visual C#/Web/Visual Studio 2012"项目，在中间部分会显示此项目下所有的项目模板列表，选择"ASP. NET MVC 4 Web 应用程序"选项。其他设置参数说明如下：

（1）.NET 框架选择：打开下拉列表框（中间上部左），选择".NET Framework 4.5.1"或其他选项；

（2）项目名称：新建项目的名称，以英文单词或汉语拼音或任意字母组合，不用汉字；

（3）位置：项目存储的本地目录位置，一般情况下，默认位置为"/项目名称/项目名称/..."；

（4）解决方案：新建解决方案，默认名称是项目名称；可另外命名，也可以选择已经存在的解决方案，也就是说，一个解决方案中可以有多个项目，解决方案是项目的容器。

2.1.3.2 项目类型选择

以上参数定义完成后,点击"确定",打开"新 ASP. NET MVC 4 项目"对话框,如图 2 - 4 所示。

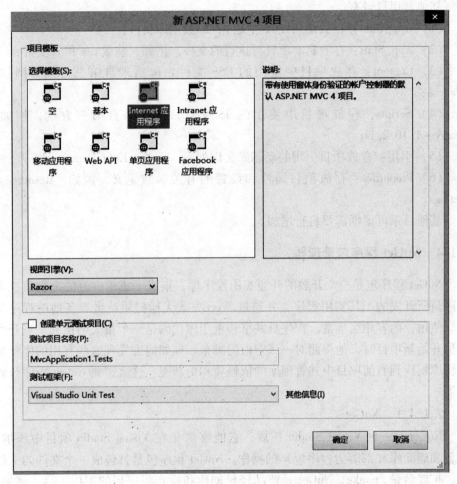

图 2 - 4 "新 ASP. NET MVC 4 项目"对话框

在左上框中选中"Internet 应用程序",其他项目使用默认选项。再次"确定",开始生成新项目。新建成的项目内容目录如图 2 - 2 所示。

2.1.3.3 项目内容目录说明

生成的项目架构是基于 MVC 框架,首先能看到以下三个目录:

(1) Controllers:存放以". cs"为扩展名的控制器(Controller)文件,例如,HomeController. cs;

(2) Models:存放以". cs"为扩展名的数据模型(Model)文件,例如,AccountViewModels. cs;

（3）Views：以控制器名称为目录名，存放相应控制器中对应方法的视图（View）文件，视图文件的扩展名为 cshtml（Razor 语法为引擎的 HTML 文件），例如，Home/Index.cshtml。

其他常用目录有：

（1）App_Data：存放持久性数据文件，即数据库文件；

（2）App_Start：存放启动项目行执行的文件，例如，装载 css 和 js 文件；

（3）Content：存放项目中使用的 CSS 文件，包括来自第三方的，例如，Site.css；

（4）Scripts：存放项目中使用的 JS 文件，包括来自第三方的，例如，jQuery-1.10.2.js；

（5）引用：存放项目引用的动态库文件列表，例如，EntityFramework；

（6）Properties：存放项目属性和设置的有关参数定义，例如，Resources、Settings。

其他目录可根据需要自行增加。

2.1.4 NuGet 程序包管理器

NuGet 程序包是一个开放的开源实用程序库，集成了成千上万的来自第三方程序员所开发的工具实用程序，并通过 Web 方式无偿提供给世界各地的程序员共享使用，内容相当丰富。程序员共享和重用库代码是一个很大的挑战，当开发人员开始新项目时，他将面对一张空白的画布，他如何去发现这些有用的库？如何将库集成到当前项目中并管理库的依赖项和更新呢？这就需要 NuGet 程序包管理器。

2.1.4.1 NuGet

NuGet 是一种 Visual Studio 扩展，它能够简化在 Visual Studio 项目中添加、更新和删除库（部署为程序包）的操作。NuGet 程序包是打包成一个文件的文件集，扩展名为.nupkg，NuGet 是产品轻松创建和发布程序包的实用工具。例如，ELMAH（Error Logging Modules and Handlers）就是一个非常有用的库，是由开发人员自己编写的。ELMAH 能够在出现异常时记录 Web 应用程序中所有未经处理的异常以及所有请求信息，例如，标头、服务器变量等。如果需要在项目中使用 ELMAH，那么就可以利用 NuGet 程序包管理器实现查询、安装，然后在项目中引用并使用其中的方法。

例如，当我们的项目里要引用第三方的库时，比如 jQuery、Newtonsoft.Json、log4net 等，我们需要从网上下载这些库，然后依次拷贝到各个项目中，当有的类库有更新时，不得不再重复一遍，这是很繁琐的事，这时就可以考虑使用 NuGet 来帮我们管理和更新这些类库，而且更新类库时会自动添加类库的相关引用，方便至极。当然网上一些我们常用的类库，更新频率不是很高，而且即便出了新版本，我们也没必要总是保持最新，故这点对我们的帮助比较有限。NuGet 最大的好处在于可以搭建自己的类库服务器，在一些规模较大的公司里面有很多的项目，然后其中有一些是整个组、甚至整个公司通用的类库，当这些类库更新后，我们需要依次拷贝到我们的项目中去，甚至有时候我们自己都搞不清楚各个项目里的版本是否一致，有时偶尔一两个项目忘了复制更新便会出现莫名其妙的错误，这令我们程序员难以处理。

2.1.4.2 安装 NuGet

点击"Tools"（工具）|"Extension Manager"（扩展和更新）菜单选项，打开 Visual Studio Extension Manager，单击"Online Gallery"（联机库）选项卡查看可用的 Visual Studio 扩展名，如图 2－5 中所示，还可以使用右上角的搜索输入框输入关键字搜索找到它，单击"Download"（下载）按钮，即可安装 NuGet（如果已经安装，则显示对勾）。

图 2－5　安装 NuGet 程序包管理器对话框

如果您已经安装了 ASP.NET MVC，在项目建立后，NuGet 程序包管理器会自动安装到项目中去。

2.1.4.3 安装程序包

这是程序员利用 NuGet 程序包管理器最常用的操作，可以用两种方式安装所需要的程序包（第三方工具）：向导和控制台。NuGet 同样内置于 Windows PowerShell 的控制台，控制台面向高级用户。

启动 NuGet，选中解决方案资源管理器所显示的项目录中的"引用"节点

（或解决方案下的项目节点），然后单击右键，在菜单中选择"管理 NuGet 程序包"（Manage NuGet Packages）选项，启动"管理 NuGet 程序包"（Manage NuGet Packages）对话框，如图 2-6 中所示。

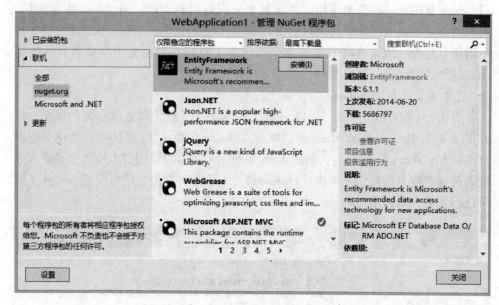

图 2-6 "管理 NuGet 程序包"对话框

左边是"已安装"、"联机"、"更新"选择，中间是对应的程序包列表，右边是对应程序包的说明。对于选中的程序包，可有三种操作："安装"、"卸载"、"更新"。找到程序包后，"安装"、"更新"完成该程序包的下载，并安装到项目引用中，包括其依赖项，并将任何必要更改应用到程序包指定的项目中；"卸载"完成选中的程序包从当前项目的引用中删除，包括其依赖项的程序包。

NuGet 安装程序包的具体步骤：

（1）下载程序包文件及其所有依赖项。有些程序包要求接受许可，并提示用户接受程序包的许可条款；大多数程序包支持隐式接受许可，并不发出提示。如果解决方案或本地计算机缓存中已经存在该程序包，NuGet 将跳过下载程序包的步骤。

（2）提取程序包的内容。NuGet 将内容提取到程序包文件夹中（在必要时创建文件夹）。程序包文件夹在您的解决方案（.sln）文件的并列位置。如果解决方案的多个项目中安装了同一个程序包，则仅提取该程序包一次并由各项目共享。

（3）引用程序包中的程序集。依照惯例，NuGet 更新项目将引用程序包 lib 文件夹中的一个或多个特定程序集。例如，将程序包安装到面向 Microsoft.NET Framework 5 的项目时，NuGet 将添加对 lib/net5 文件夹中的程序集的引用。

(4)将内容复制到项目中。依照惯例，NuGet将程序包的内容文件夹的内容复制到项目中，这对包含JavaScript文件或图像的程序包十分有用。

(5)应用程序包转换。如果任何程序包包含转换文件，例如用于配置的App.config.transform或Web.config.transform，则NuGet将在复制内容之前应用这些转换。有些程序包所含的源代码经过转换可以在源文件中包含当前项目的命名空间，NuGet将同样转换这些文件。

(6)运行程序包中关联的Windows PowerShell脚本。有些程序包可能包含Windows PowerShell脚本，这些脚本使用设计时环境(DTE)自动化Visual Studio，从而处理与NuGet无关的任务。

当NuGet执行所有这些步骤后，库就准备就绪。很多程序包使用WebActivator程序包自行激活，从而最小化安装后所需的任何配置。

程序包不再需要时可以卸载，使项目回到安装程序包之前的状态。

2.1.4.4 面向高级用户的NuGet

程序员可以使用命令行shell命令完成同样的任务，控制台窗口（程序包管理器控制台）以及一组Windows PowerShell命令与NuGet进行交互。

启动程序包管理器控制台："工具"（Tools）｜"NuGet程序包管理器"（Library Package Manager）｜"程序包管理器控制台"（Package Manager Console）菜单选项，如图2-7所示。

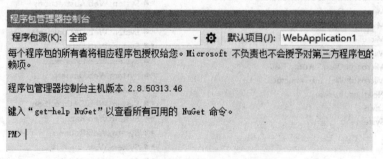

图2-7 NuGet程序包安装控制台

控制台使用命令：

(1)列出程序包（Get-Package），并通过指定ListAvailable标记和Filter标记联机搜索程序包。下列命令行搜索所有包含"MVC"的程序包：

Get-Package-ListAvailable-Filter Mvc

(2)安装程序包（Install-Package），例如，将ELMAH安装到当前项目的命令：

Install-Package Elmah

(3)更新程序包（Update-Package），程序包管理器控制台还包含一个命令，与对话框相比，它提供更多的更新控制。例如，无需参数即可调用此命令以更新解决方案的每个项目中的各个程序包：

Update – Package

此命令尝试将每个程序包都更新到最新版本。因此，如果您有 1.0 版本的程序包，而 1.1 和 2.0 版本在该程序包源中可用，则该命令将此程序包更新至最新的 2.0 版本。

有关 NuGet 程序包管理器的更多功能请参考有关手册。

2.1.5 引用目录内容

Web MVC 项目建立后，项目中自动生成一个名为"引用（References）"的目录，如图 2-8 所示，其中存放的是项目设计和运行时所必需的扩展库。

```
▲ ■ 引用
    ■ Antlr3.Runtime
    ■ EntityFramework
    ■ EntityFramework.SqlServer
    ■ Microsoft.AspNet.Identity.Core
    ■ Microsoft.AspNet.Identity.EntityFramework
    ■ Microsoft.AspNet.Identity.Owin
    ■ Microsoft.CSharp
    ■ Microsoft.Owin
    ■ Microsoft.Owin.Host.SystemWeb
    ■ Microsoft.Owin.Security
    ■ Microsoft.Owin.Security.Cookies
    ■ Microsoft.Owin.Security.Facebook
    ■ Microsoft.Owin.Security.Google
    ■ Microsoft.Owin.Security.MicrosoftAccount
    ■ Microsoft.Owin.Security.OAuth
    ■ Microsoft.Owin.Security.Twitter
    ■ Microsoft.Web.Infrastructure
    ■ Newtonsoft.Json
    ■ Owin
    ■ System
    ■ System.ComponentModel.DataAnnotations
    ■ System.Configuration
    ■ System.Core
    ■ System.Data
    ■ System.Data.DataSetExtensions
    ■ System.Drawing
    ■ System.EnterpriseServices
    ■ System.Net.Http
    ■ System.Net.Http.WebRequest
    ■ System.Web
    ■ System.Web.Abstractions
    ■ System.Web.ApplicationServices
    ■ System.Web.DynamicData
    ■ System.Web.Entity
    ■ System.Web.Extensions
    ■ System.Web.Helpers
    ■ System.Web.Mvc
    ■ System.Web.Optimization
    ■ System.Web.Razor
    ■ System.Web.Routing
    ■ System.Web.Services
    ■ System.Web.WebPages
    ■ System.Web.WebPages.Deployment
    ■ System.Web.WebPages.Razor
    ■ System.Xml
    ■ System.Xml.Linq
    ■ WebGrease
```

图 2-8 引用目录内容

这些动态程序库有些是通过 Web.config 配置设置直接引用，例如，System.Web.Helpers；有些需要显式引用（在代码中使用"using xxx"，例如，System.Web.Mvc）命名空间及其子项目。查看某个引用库的内容及包含的方法可以使用右键菜单。

2.2 建立监理信息系统项目

"建设工程监理信息系统"项目名称定义为"psjlmvc4",项目存放于 E:\(E 盘根目录)。项目设计开发完成后的内容目录如图 2-9 所示。

图 2-9　psjlmvc4 项目内容目录

各目录内容及作用下面分别说明。

2.2.1 项目属性（Properties）

项目属性参数定义分为"应用程序"、"生成"、"Web"、"打包/发布 Web"、"打包/发布 SQL"、"Silverlight 应用程序"、"生成事件"、"资源"、"设置"、"引用路径"、"签名"、"代码分析" 12 个类别。

2.2.1.1 应用程序

应用程序类别包含有关项目的常规参数设置，例如，程序集名称、命名空间、目标框架、应用程序显示图标等，另外，通过"程序集信息"还可以设置与程序项目有关的项目名称、公司名称、版本号等参数，如图 2-10 和图 2-11 所示。

其中，应用程序集名称和默认名空间名称使用项目名称，即"psjlmvc4"；目标架构设置为".NET Framework 4.5"或".NET Framework 4.5.1"。

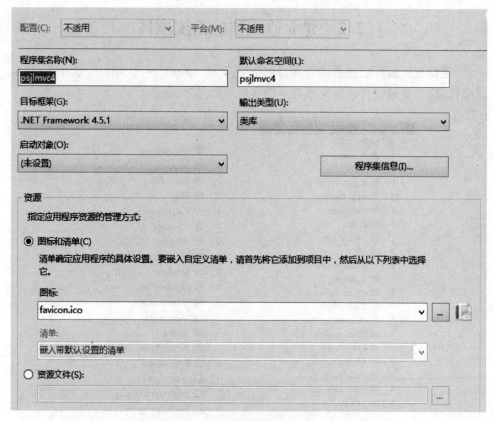

图 2-10 psjlmvc4 项目属性—应用程序设置

2.2.1.2 生成

生成是项目编译成目标运行代码后的运行平台参数，一般情况下采用系统默认值，例如，配置＝活动（Debug）、平台＝活动（Any CPU）、输出路径＝\BIN。

2.2.1.3 Web

Web 类别参数是指项目调试运行时的 Web 支持服务器，默认值是"IIS Express"，可选"本地 IIS"，利用实际运行环境调试项目的运行情况；项目 URL（虚拟目录）设置为"http://localhost/psjlmvc4"；调试器选择为"ASP.NET"，可同时选择的选项还有"本机代码"、"SQL Server"、"Silverlight"。

2.2.1.4 打包/发布 Web

项目完成、调试成功后，就可以部署应用程序项目到其他服务器，使其成为可以通过 Internet 访问的应用程序。有一键式发布和 Web 部署软件包部署 Web 应用程序项目。

要部署的项目（适用于所有部署方法）：仅限运行此应用程序所需要的文件，其他选项还有"此项目中的所有文件"、"此项目文件夹中的所有文件"。

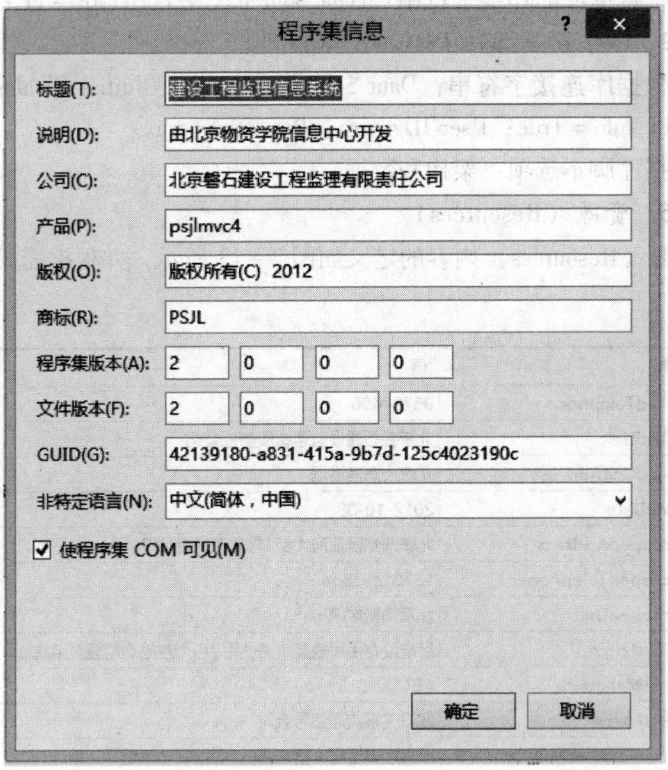

图 2-11　psjlmvc4 项目属性—应用程序集参数设置

2.2.1.5　打包/发布 SQL

有关数据库随 Web 项目内容同时发布的参数设置，可以通过 FTP 等方式实现数据库的上传更新，采用默认值（不设置），否则，相应属性设置如下：

（1）数据库项：选择"从 Web.config 导入"，结果如图 2-12 所示，其中第一个是本项目实际使用的数据库连接字符名称，后两个是项目自定义连接字符串。

图 2-12　psjlmvc4 项目属性—打包/发布 SQL

（2）目标数据库的连接字符串：Data Source = WGXHOME – PC；Initial Catalog = psjldb12；User ID = wgx；Password = wgx。

（3）源数据库连接字符串：Data Source =（local）；Initial Catalog = psjldb12；Persist Security Info = True；User ID = wgx；Password = wgx。

（4）数据库脚本选项：架构和数据。

2.2.1.6 资源（Resources）

系统资源（Resources）内容的定义如图2-13所示，并根据需要增加必要的项目。

名称	值	注释
AppliedTelephone	86798456	
AppliedUnit	北京磐石建设监理有限责任公司	
AppliedUnitAddress	北京市海滨浴场	
CreateDate	2012-10-08	
DevolopedAddress	北京通州区富河大街1号北京物资学院	
DevolopedTelephone	13701397654	
DevolopedUnit	北京物资学院	
NoUserLogin	目前没有用户登录！请使用[用户切换]功能登录系统…	
ProjectManager	ABCD	
ProjectName	建设工程监理信息系统	
ProjectNotSelected	没有选择工程！请使用[工程选择]功能选择工程…	
VersionCode	2.0	

图2-13 系统资源属性及定义

2.2.1.7 设置（Settings）

系统设置（Settings）内容的定义如图2-14所示，并根据需要增加必要的项目。

名称	类型	范围	值
Projectcode	string	应用程序	2012-0001
Supervisor	string	应用程序	ABCD
CreateDate	System.Da…	应用程序	2012-10-07 12:32

图2-14 系统设置（Settings）内容的定义

其他类别属性值采用系统默认值。

2.2.2 区域目录（Areas）

"Areas"是系统约定的命名空间（目录）名称，意为"区域"，是分隔大型项目资源存放管理的一种方法。根据此思想，本项目根据功能分为若干个 Area（区域）分别存储相应的文件资源，进行分类管理，并用大写英文字母依次标记各个子区域的名称，例如，"工程项目"管理功能区域目录名为"AArea"、"文档管理"功能目录区域名称为"BArea"，以此类推，共有 A～U 20 个区域，如图 2－15 所示。

图 2－15 Areas 目录内容

2.2.2.1 区域目录与系统功能

区域目录的划分是根据功能类别为基础，其作用是存放相应功能所需要的运行文件和资源文件。表 2－1 列出了本项目管理功能名称与管理功能资源存放区（命名空间、目录）的对照表。

表 2－1 管理功能名称与管理功能资源存放区域对照表

功　　能	区　　域	说　　明
A－工程项目	AArea	工程信息管理
B－文档管理	BArea	监理档案管理
C－前期准备	CArea	监理前期准备
D－施工准备	DArea	施工准备阶段
E－进度控制	EArea	工程进度控制
F－质量控制	FArea	工程质量控制
G－造价控制	GArea	工程造价控制
H－施工合同其他事项	HArea	施工合同其他项目
I－规程法规	IArea	政策法律法规文件
J－查询统计	JArea	数据查询统计
K－系统管理	KArea	系统设置管理
L－基础数据	LArea	配套常用数据
M－安全管理	MArea	安全监理管理
O－其他辅助信息管理	OArea	辅助功能管理
┆	┆	增加新项目、新功能

2.2.2.2 区域（目录）结构

每个区域（目录）的结构要严格按照 MVC 架构要求的结构组成，即由"Controllers"、"Models"、"Views" 和必要的文件，以下以 AArea 为例加以说明。

AArea 子目录中的内容由以下 4 部分组成：

（1）Controllers：控制器目录，存放相应的控制器文件；

（2）Models：模型目录，存放相应的模型文件；

（3）Views：视图目录，存放对应控制器中方法的视图文件，其中文件"_ViewStart.cshtml"与主 MVC 视图目录的同名文件内容相同；

（4）AAreaAreaRegistration.cs：区域注册和区域路由注册定义文件。

2.2.2.3 区域注册与路由注册定义文件

"AAreaAreaRegistration.cs"是用户在新建区域目录时自动生成的配置参数定义文件，称为区域注册和区域路由注册定义文件，其功能是完成区域路由导航，内容是建立区域时自动生成的，无需更改。其内容如下：

```
using System.Web.Mvc;
namespace psjlmvc4.Areas.AArea
{
    public class AAreaAreaRegistration:AreaRegistration
    {
        public override string AreaName
        {
            get
            {
                return"AArea";
            }
        }
        public override void RegisterArea(AreaRegistrationContext context)
        {
            context.MapRoute(
                "AArea_default",
                "AArea/{controller}/{action}/{id}",
                new{action="Index",id=UrlParameter.Optional}
            );
        }
    }
}
```

2.3 MVC 目录架构

MVC 目录架构由三个目录（命名空间）组成，即"Controllers"、"Models"、"Views"，这是 MVC 项目的主体目录结构。

2.3.1 控制器目录（Controllers）

控制器目录存放相应的控制器文件，控制器文件的类型为".cs"，即 C#类文件。本项目控制器文件有：

（1）AccountController.cs：VS2013 账户管理控制器（VS2013 系统自定义，不作讨论）；

（2）HomeController.cs：项目入口主控制器，包括项目主页（Index）、用户登录（Login）、关于我们（About）、菜单显示（SubMenuPartial）等方法；

（3）OtherController.cs：其他控制器，包括图片处理、图片显示、系统信息显示等方法。

其他控制器文件分别存储于对应功能区域子目录中的"Controllers"目录中。

2.3.2 模型目录（Models）

此处的模型目录存放项目所使用的分类数据模型文件，模型文件的类型为".cs"，即 C#类文件。本项目公共模型文件目录有：

（1）AAppendModels.cs：附录 A 对应表格的模型集合；

（2）AccountModels.cs：VS2013 账户管理用模型；

（3）AppService.cs：应用服务类，包括数据读取、存储、金额变换等自定义方法；

（4）BAppendModels.cs：附录 B 对应表格的模型集合；

（5）CAppendModels.cs：附录 C 对应表格的模型集合；

（6）CommonModels.cs：公共模型，包括模型信息记录模型、系统信息记录模型等；

（7）DAppendModels.cs：附录 D 对应表格的模型集合；

（8）EAppendModels.cs：附录 E 对应表格的模型集合；

（9）EntityAttribute.cs：实体属性管理类，包括实体属性的读取方法；

（10）psjldbContext.cs：数据库连接上下文类（继承自 DbContext 类），内容是项目所有模型的连接对应表名列表，例如，public DbSet < AProjectList > AProjectLists ｛get; set;｝，即工程项目信息记录模型（AProjectList）对应的数据库表名是 AProjectLists（其命名规则遵照系统约定优先规则）。

2.3.3 视图目录（Views）

视图目录存放控制器对应方法处理结果（数据）的客户端，展示所需要的前端 HTML 文件，其内容结构如图 2-16 所示，其中子目录的名称根据对应控制器的名称（主名称），并在生成控制器时自动生成，相应方法对应的 HTML 文件存放其中；其中另外的两个文件是"_ViewStart.cshtml"和"Web.config"。

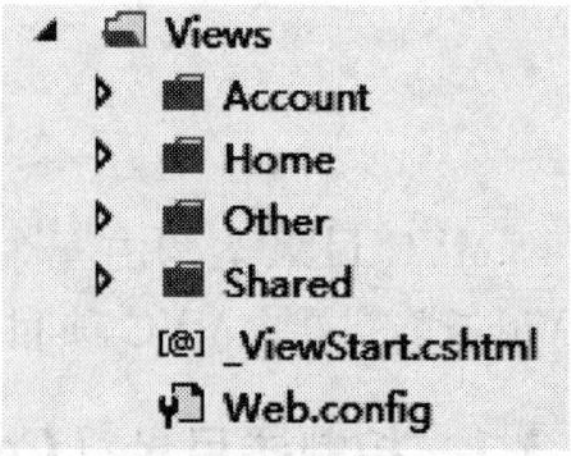

子目录名是根据约定规则自动生成的，例如，子目录"Home"对应于控制器"HomeController.cs"、子目录"Other"对应于控制器"OtherController.cs"。

图 2-16 视图目录的内容

每个子目录中的内容是相应控制器中对应方法（Action）的视图文件，以控制器"HomeController.cs"为例，其中包括的方法（Action）如下：

```
namespace psjlmvc4.Controllers
{
    public partial class HomeController:Controller
    {
        public virtual ActionResult Index(string funcode = null, string mtitle = null)
        {……;return View(rd.ToList());}
        public virtual ActionResult About()
        {……;return View();}
        public virtual ActionResult Contact()
        {……;return View();}
        public virtual ActionResult Loging(string returnUrl)
        {……;return View();}
        public virtual ActionResult LogOff(bool qexit = false)
        {……;}
        public ActionResult SubMenuPartial(string fcode = null)
        {……;return PartialView(rd.ToList());}
        public virtual ActionResult Updatepwd(KUserAdd up)
        {……;return View(up);}
        public virtual ActionResult pSelect(FormCollection fc)
        {……;return View(rd.ToList());}
        public virtual ActionResult pSelected(string pcode, string pname, string pdate)
        {……;return RedirectToAction("Index","Home");}
        private int OnlineCount(string uid = "", bool ps = false)
        {……;return db.OnlineUsers.Count();}
        public ActionResult Layout14()
        {……;return View();}
        protected override void Dispose(bool disposing)
        {……;}
    }
}
```

"ActionResult"标识符之后的标识符对应一个方法（Action），如果处理结果（数据）需要展示给客户端，那么就需要一个对应的视图（.cshtml），否则可以没有对应的视图。上述控制器对应的视图子目录中的内容如图2-17所示（以字母顺序排列）。

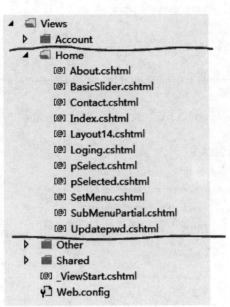

图2-17 HomeController.cs 对应的视图目录内容

对于方法"pSelected"、"Dispose"、"LogOff"则没有对应的方法视图，因为其任务是后台处理或重定向到其他方法视图。

"Shared"是视图共享目录，存放公用的视图，例如，项目视图模板文件"_Layout.cshtml"，运行状态信息显示视图"Error.cshtml"。此目录中的视图文件访问不受路由规则限制，因为"Shared"是优先访问路径。

"_ViewStart.cshtml"是视图运行之前首先被访问的视图，用于存放共用模板视图文件，例如，视图模板"_Layout.cshtml"。此时"_ViewStart.cshtml"文件内容及格式如下：

```
@{
    Layout = "~/Views/Shared/_Layout.cshtml";
}
```

"Web.config"是视图设计、运行的架构信息配置文件。

2.3.4 路由规则定义文件

MVC架构的目录名称命名遵照"系统约定"规则，符合"RouteConfig.cs"

路由配置文件中定义的路由访问规则。

"RouteConfig.cs" 文件位于目录 "App_Start" 中，其内容如下：

```csharp
using System.Web.Mvc;
using System.Web.Routing;namespace psjlmvc4
{
    public class RouteConfig
    {
        public static void RegisterRoutes(RouteCollection routes)
        {
            routes.IgnoreRoute("{resource}.axd/{*pathInfo}");
            routes.MapRoute(
                "Default",
                "{controller}/{action}/{id}",
                new { controller = "Home", action = "Index",
                    id = UrlParameter.Optional },
                new string[] {"psjlmvc4.Controllers"}
            );
        }
    }
}
```

2.3.5 其他目录说明

其他目录，由用户自定义生成和系统约定生成两种方式产生。

2.3.5.1 系统约定生成的目录

Content：存放 CSS 类文件。这些文件来自于自定义或随 jQuery 控件一起的配套层叠样式文件，本项目使用的 CSS 文件有 "psjlmvc4.css"、"WebGrid.css"、"PagedList.css"、"bootstrap.css" 等，另外子目录 "Theme" 中存放的 CSS 文件是 jQuery 控件一起的配套层叠样式文件，可以单独使用。

App_Data：ASP.NET 系统目录，存放数据库文件，本项目使用的数据库文件是 "psjldb12.mdf"。

App_Start：ASP.NET 系统目录，存放项目运行之前需要先行执行的文件，例如，权限认证配置文件 "AuthConfig.cs"、CSS 和 JS 引入绑定文件 "BundleConfig.cs"、路由规则定义文件 "RouteConfig.cs" 等。

CodeTemplates：ASP.NET 系统目录，代码生成模板文件，包括 MVC 控制器代码生成模板和 CRUD 视图 CSHTML 代码生成模板，模板文件以 ".tt" 为扩展名。

obj：ASP. NET 系统目录，系统编译时的文件临时存储目录。
bin：ASP. NET 系统目录，系统编译后的动态链接库（dll）文件存储目录。
Scripts：ASP. NET 系统目录，是比较重要的目录，存放项目所使用的 JS 文件，或者是自定义的 JS 文件，或者是随 jQuery 一起配套使用的 JS 文件。

2.3.5.2 用户自定义目录

BasicSlider：自定义目录，存放图片显示控件所使用的图片、CSS 和 JS 文件。
ContractFiles：自定义目录，存放合同原始文件的扫描件或图片文件（原稿）。
DownLoadFiles：自定义目录，存放各种可供远程客户端下载的文件。
DrawingFiles：自定义目录，存放工程施工有关的图纸扫描件或图片文件。
Extensions：自定义目录，存放自定义的应用方法扩展文件。
Images：ASP. NET 系统目录，新建项目时自动生成的用于存放图片文件的目录。
ImageFiles：自定义目录，自定义的用于存放图片文件的目录。
MusicFiles：自定义目录，存放音乐文件。
PlayImages：自定义目录，存放两种图片轮播控件所需要的 CSS 和 JS 文件，一种是"NivoSlider"，另一种是"SlideBox"。
StudyFiles：自定义目录，存放用于系统使用说明和用户学习材料所需要的文件。
Unslider：自定义目录，存放图片轮播控件"Unslider"所需要的 CSS、JS 和相应的图片文件。
UploadFiles：自定义目录，存放远程客户端上传的文件。
UserImages：自定义目录，存放系统用户头像文件（图片）。
VideoFiles：自定义目录，存放视频文件，包括动画文件。

2.4　Web. config 文件

Web. config 是 Web 应用项目开发设计和运行时所需要的配置参数定义文件。因为除了项目根目录外，其他目录也可能有此文件，因此，此处所讨论的 Web. config 是指项目根目录中的 Web. config 文件，其他目录中的 Web. config 文件内容是根目录中 Web. config 文件内容的子集。

2.4.1　Web. config 文件结构说明

通过 VS2013 新建一个 Web 应用项目后，默认情况下会在根目录自动创建一个 Web. config 文件，其内容包括了系统的默认设置。Web. config 配置文件（默认的配置设置）的内容结构如下：

```
<?xml version="1.0"?>
<configuration>
    ……
    <system.web>
        ……
    </system.web>
    ……
</configuration>
```

文件内容以语句"<?xml version="1.0"?>"开始，说明"Web.config"文件内容的语法是 XML 格式。

文件内容结构以标识符"<configuration>"开始，以"</configuration>"标识符结束，中间内容以"<XXX>……</XXX>"为标记的若干节（Section），每节还可以有自己的子节（Subsection）。

2.4.2 主要节功能说明

2.4.2.1 <authentication>节

作用：配置 ASP.NET 身份验证支持（为 Windows、Forms、PassPort、None4种），该元素只能在计算机、站点或应用程序级别声明。<authentication>元素必须与<authorization>节配合使用。

以下示例为基于窗体（Forms）的身份验证配置站点，当未登录的用户访问需要身份验证的网页时，网页将自动跳转到登录网页。

```
<authentication mode="Forms">
    <forms loginUrl="logon.aspx" name=".FormsAuthCookie"/>
</authentication>
```

loginUrl：表示登录网页的名称。
name：表示 Cookie 名称。

2.4.2.2 <authorization>节

作用：控制 URL 资源的客户端访问（如允许匿名用户访问）。此元素可以在任何级别（计算机、站点、应用程序、子目录或页）上声明；必须与<authentication>节配合使用。

下面来看一个禁止匿名用户访问的示例。

```
< authorization >
    < deny users = "?"/ >
</ authorization >
```

提示：可以使用 user. identity. name 来获取已验证的当前用户名；可以使用 web. Security. FormsAuthentication. RedirectFromLoginPage 方法将已验证的用户重新定向到用户刚才请求的页面。

2.4.2.3 < compilation > 节

作用：配置 ASP. NET 使用的所有编译设置。默认的 debug 属性为 True，在程序编译完成交付使用之后应将其设为 True（Web. config 文件中有详细说明，此处省略示例）。

2.4.2.4 < customErrors > 节

作用：为 ASP. NET 应用程序提供有关自定义错误的信息（不适用于 XML Web Services 中发生的错误）。

示例：当发生错误时，将网页跳转到自定义的错误页面。

```
< customErrors defaultRedirect = "ErrorPage. aspx" mode = "RemoteOnly" >
</ customErrors >
```

defaultRedirect：表示自定义的错误网页的名称。

mode：元素表示对不在本地 Web 服务器上运行的用户显示自定义（友好的）信息。

2.4.2.5 < httpRuntime > 节

作用：配置 ASP. NET HTTP 运行库设置。该节可以在计算机、站点、应用程序和子目录级别声明。

示例：控制用户上传文件最大为 4MB，最长时间为 60 秒，最多请求数为 100。

```
< httpRuntime maxRequestLength = "4096"
executionTimeout = "60" appRequestQueueLimit = "100"/ >
```

2.4.2.6 < pages > 节

作用：该标识用于特别指定页的配置（如是否启用会话状态、视图状态，是否检测用户的输入等）。< pages > 可以在计算机、站点、应用程序和子目录级别声明。

示例：不检测用户在浏览器输入的内容中是否存在潜在的危险数据（注意：该项默认为检测，如果使用了不检测，一定要对用户的输入进行编码或验证），在从客户端发回页时将检查加密的视图状态，以验证视图状态是否已在客户端被篡改。

```
< pages buffer = "true" enableViewStateMac = "true" validateRequest = "false"/ >
```

2.4.2.7 < sessionState > 节

作用：为当前应用程序配置会话状态设置（如设置是否启用会话状态、会话状态保存位置）。示例如下：

```
< sessionState mode = "InProc" cookieless = "true" timeout = "20"/ >
</sessionState >
```

mode = "InProc"：表示在本地存储会话状态（可以选择存储在远程服务器或 SAL 服务器中或不启用会话状态）。

cookieless = "true"：表示如果用户浏览器不支持 Cookie，则启用会话状态（默认为 false）。

timeout = "20"：表示会话可以处于空闲状态的分钟数。

2.4.2.8 < trace > 节

作用：配置 ASP.NET 跟踪服务，主要用来测试判断程序哪里出错。示例：以下为 Web.config 中的默认配置。

```
< trace enabled = "false" requestLimit = "10" pageOutput = "false" traceMode = "SortByTime" localOnly = "true"/ >
```

enabled = "false"：表示不启用跟踪。

requestLimit = "10"：表示指定在服务器上存储的跟踪请求的数目。

pageOutput = "false"：表示只能通过跟踪实用工具访问跟踪输出。

traceMode = "SortByTime"：表示以处理跟踪的顺序来显示跟踪信息。

localOnly = "true"：表示跟踪查看器（trace.axd）只用于宿主 Web 服务器。

2.4.3 Web.config 文件内容示例

```
<? xml version = "1.0"standalone = "yes"? >
< configuration xmlns = "http://schemas.microsoft.com/.NetConfiguration/v2.0" >
    <! --配置全局变量 -- >
```

```
< appSettings >
    < add key = "examstr"
value = "server = . ;database = myweb_exam_db;uid = exam_login;pwd = xxd_examadmin"/ >
</ appSettings >
<！--网站系统配置-- >
< system. web >
    <！--上传文件时提示访问被拒绝,生成图片文件等失败文件夹没有改写权限-- >
    < identity impersonate = "true"/ >
    <！--在客户端显示错误信息-- >
    < customErrors mode = "Off"/ >
    <！--允许匿名访问-- >
    < authorization >
        < allow users = " * "/ >
    </authorization >
    <！--启用跟踪页面-- >
    < trace enabled = "true" requestLimit = "1000" pageOutput = "true"
traceMode = "SortByTime"localOnly = "true"/ >
    <！--防止网页乱码-- >
    < globalization requestEncoding = "gb2312" responseEncoding = "gb2312"
culture = "zh - CN"fileEncoding = "gb2312"/ >
    <！--有两个form时,用窗体提交时配置-- >
    < pages validateRequest = "false" enableSessionState = "true"enableViewState = "true"
enableEventValidation = "false"/ >
    <！--设置上传文件大小-- >
    < httpRuntime executionTimeout = "300" maxRequestLength = "102400"
useFullyQualifiedRedirectUrl = "false"/ >
</ system. web >
</ configuration >
```

2.4.4 本项目Web. config文件内容

```
<？xml version = "1. 0"？ >
< configuration >
    < configSections >
        < section name = "entityFramework"
type = "System. Data. Entity. Internal. ConfigFile. EntityFrameworkSection,EntityFramework,
Version = 5. 0. 0. 0,Culture = neutral,PublicKeyToken = b77a5c561934e089"
requirePermission = "false"/ >
```

```xml
<sectionGroup name = "applicationSettings"
    type = "System.Configuration.ApplicationSettingsGroup, System, Version = 4.0.0.0, Culture = neutral, PublicKeyToken = b77a5c561934e089" >
    <section name = "psjlmvc4.Properties.Settings"
        type = "System.Configuration.ClientSettingsSection, System, Version = 4.0.0.0, Culture = neutral, PublicKeyToken = b77a5c561934e089" requirePermission = "false"/ >
</sectionGroup>
<sectionGroup name = "dotNetOpenAuth"
    type = "DotNetOpenAuth.Configuration.DotNetOpenAuthSection, DotNetOpenAuth.Core" >
    <section name = "messaging"
        type = "DotNetOpenAuth.Configuration.MessagingElement, DotNetOpenAuth.Core"
        requirePermission = "false" allowLocation = "true"/ >
    <section name = "reporting"
        type = "DotNetOpenAuth.Configuration.ReportingElement, DotNetOpenAuth.Core"
        requirePermission = "false" allowLocation = "true"/ >
    <section name = "openid" type = "DotNetOpenAuth.Configuration.OpenIdElement, DotNetOpenAuth.OpenId" requirePermission = "false" allowLocation = "true"/ >
    <section name = "oauth" type = "DotNetOpenAuth.Configuration.OAuthElement, DotNetOpenAuth.OAuth" requirePermission = "false" allowLocation = "true"/ >
</sectionGroup>
</configSections>
<connectionStrings>
    <add name = "psjldb12Context" connectionString = "Data Source = (local); Initial Catalog = psjldb12; Persist Security Info = True; User ID = wgx; Password = wgx; MultipleActiveResultSets = True"
        providerName = "System.Data.SqlClient"/ >
    <add name = "DefaultConnection"
        connectionString = "DefaultConnection_ConnectionString"
        providerName = "System.Data.SqlClient"/ >
    <add name = "psjlmvc4.Models.UsersContext"
        connectionString = "psjlmvc4.Models.UsersContext_ConnectionString"
        providerName = "System.Data.SqlClient"/ >
</connectionStrings>
<appSettings>
    <add key = "webpages:Version" value = "2.0.0.0"/ >
    <add key = "webpages:Enabled" value = "false"/ >
    <add key = "PreserveLoginUrl" value = "true"/ >
    <add key = "ClientValidationEnabled" value = "true"/ >
    <add key = "UnobtrusiveJavaScriptEnabled" value = "true"/ >
</appSettings>
```

```
<system.web>
    <compilation targetFramework = "4.5.1" debug = "true"/>
    <httpRuntime targetFramework = "4.5"/>
    <authentication mode = "Forms">
        <forms loginUrl = "~/Home/Loging" timeout = "100" protection = "All">
            <credentials passwordFormat = "SHA1"/>
        </forms>
    </authentication>
    <authorization>
        <allow users = "*"/>
        <deny users = "?"/>
    </authorization>
    <sessionState mode = "InProc" timeout = "60"/>
    <pages>
            <namespaces>
                <add namespace = "System.Web.Helpers"/>
                <add namespace = "System.Web.Mvc"/>
                <add namespace = "System.Web.Mvc.Ajax"/>
                <add namespace = "System.Web.Mvc.Html"/>
                <add namespace = "System.Web.Optimization"/>
                <add namespace = "System.Web.Routing"/>
                <add namespace = "System.Web.WebPages"/>
            </namespaces>
            <controls>
                <add tagPrefix = "c" namespace = "Web.Ajax.Controls" assembly = "Web.Ajax"/>
                <add tagPrefix = "ajaxToolkit" assembly = "AjaxControlToolkit" namespace = "AjaxControlToolkit"/>
            </controls>
        </pages>
        <identity impersonate = "true"/>
        <httpHandlers>
            <remove verb = "*" path = "resource.axd"/>
            <add verb = "GET,HEAD" path = "resource.axd" type = "Web.Ajax.Handlers.Resource" validate = "true"/>
        </httpHandlers>
</system.web>
    <runtime>
        <assemblyBinding xmlns = "urn:schemas-microsoft-com:asm.v1">
            <dependentAssembly>
```

```xml
            <assemblyIdentity name = "System.Web.Helpers"
publicKeyToken = "31bf3856ad364e35"/>
                <bindingRedirect oldVersion = "1.0.0.0-2.0.0.0" newVersion = "2.0.0.0"/>
            </dependentAssembly>
            <dependentAssembly>
                <assemblyIdentity name = "System.Web.Mvc"
publicKeyToken = "31bf3856ad364e35"/>
                <bindingRedirect oldVersion = "0.0.0.0-4.0.0.0" newVersion = "4.0.0.0"/>
            </dependentAssembly>
            <dependentAssembly>
                <assemblyIdentity name = "System.Web.WebPages"
publicKeyToken = "31bf3856ad364e35"/>
                <bindingRedirect oldVersion = "0.0.0.0-2.0.0.0" newVersion = "2.0.0.0"/>
            </dependentAssembly>
            <dependentAssembly>
                <assemblyIdentity name = "DotNetOpenAuth.AspNet"
publicKeyToken = "2780ccd10d57b246" culture = "neutral"/>
                <bindingRedirect oldVersion = "0.0.0.0-4.3.0.0" newVersion = "4.3.0.0"/>
            </dependentAssembly>
            <dependentAssembly>
                <assemblyIdentity name = "DotNetOpenAuth.Core"
publicKeyToken = "2780ccd10d57b246" culture = "neutral"/>
                <bindingRedirect oldVersion = "0.0.0.0-4.3.0.0" newVersion = "4.3.0.0"/>
            </dependentAssembly>
            <dependentAssembly>
                <assemblyIdentity name = "WebGrease"
publicKeyToken = "31bf3856ad364e35" culture = "neutral"/>
                <bindingRedirect oldVersion = "0.0.0.0-1.5.2.14234"
newVersion = "1.5.2.14234"/>
            </dependentAssembly>
            <dependentAssembly>
                <assemblyIdentity name = "DotNetOpenAuth.OpenId"
publicKeyToken = "2780ccd10d57b246" culture = "neutral"/>
                <bindingRedirect oldVersion = "0.0.0.0-4.1.0.0" newVersion = "4.1.0.0"/>
            </dependentAssembly>
            <dependentAssembly>
                <assemblyIdentity name = "DotNetOpenAuth.OAuth"
publicKeyToken = "2780ccd10d57b246" culture = "neutral"/>
                <bindingRedirect oldVersion = "0.0.0.0-4.1.0.0" newVersion = "4.1.0.0"/>
            </dependentAssembly>
```

```xml
<dependentAssembly>
    <assemblyIdentity name="DotNetOpenAuth.OAuth.Consumer" publicKeyToken="2780ccd10d57b246" culture="neutral"/>
    <bindingRedirect oldVersion="0.0.0.0-4.1.0.0" newVersion="4.1.0.0"/>
</dependentAssembly>
<dependentAssembly>
    <assemblyIdentity name="DotNetOpenAuth.OpenId.RelyingParty" publicKeyToken="2780ccd10d57b246" culture="neutral"/>
    <bindingRedirect oldVersion="0.0.0.0-4.1.0.0" newVersion="4.1.0.0"/>
</dependentAssembly>
<dependentAssembly>
    <assemblyIdentity name="NuGet.Core" publicKeyToken="31bf3856ad364e35" culture="neutral"/>
    <bindingRedirect oldVersion="0.0.0.0-2.6.40619.9041" newVersion="2.6.40619.9041"/>
</dependentAssembly>
<dependentAssembly>
    <assemblyIdentity name="PagedList" publicKeyToken="abbb863e9397c5e1" culture="neutral"/>
    <bindingRedirect oldVersion="0.0.0.0-1.16.0.0" newVersion="1.16.0.0"/>
</dependentAssembly>
<dependentAssembly>
    <assemblyIdentity name="Microsoft.Web.XmlTransform" publicKeyToken="b03f5f7f11d50a3a" culture="neutral"/>
    <bindingRedirect oldVersion="0.0.0.0-1.2.0.0" newVersion="1.2.0.0"/>
</dependentAssembly>
</assemblyBinding>
<legacyHMACWarning enabled="0"/>
</runtime>
<entityFramework>
    <defaultConnectionFactory type="System.Data.Entity.Infrastructure.LocalDbConnectionFactory, EntityFramework">
        <parameters>
            <parameter value="v11.0"/>
        </parameters>
    </defaultConnectionFactory>
</entityFramework>
<system.webServer>
    <handlers>
```

```
            < add name = "ResourceHandler"path = "resource. axd"verb = " * "
type = "Web. Ajax. Handlers. Resource, Web. Ajax"resourceType = "Unspecified"
preCondition = "integratedMode"/ >
            < remove name = "ExtensionlessUrlHandler - ISAPI - 4. 0_32bit"/ >
            < remove name = "ExtensionlessUrlHandler - ISAPI - 4. 0_64bit"/ >
            < remove name = "ExtensionlessUrlHandler - Integrated - 4. 0"/ >
            < add name = "ExtensionlessUrlHandler - ISAPI - 4. 0_32bit"path = " * . "
verb = "GET,HEAD,POST,DEBUG,PUT,DELETE,PATCH,OPTIONS"modules = "IsapiModule"
scriptProcessor = "% windir% \Microsoft. NET\Framework\v4. 0. 30319\aspnet_isapi. dll"
preCondition = "classicMode,runtimeVersionv4. 0,bitness32"responseBufferLimit = "0"/ >
            < add name = "ExtensionlessUrlHandler - ISAPI - 4. 0_64bit"path = " * . "
verb = "GET,HEAD,POST,DEBUG,PUT,DELETE,PATCH,OPTIONS"modules = "IsapiModule"
scriptProcessor = "% windir% \Microsoft. NET\Framework64\v4. 0. 30319\aspnet_isapi. dll"
preCondition = "classicMode,runtimeVersionv4. 0,bitness64"responseBufferLimit = "0"/ >
            < add name = "ExtensionlessUrlHandler - Integrated - 4. 0"path = " * . "
verb = "GET,HEAD,POST,DEBUG,PUT,DELETE,PATCH,OPTIONS"
type = "System. Web. Handlers. TransferRequestHandler"
preCondition = "integratedMode,runtimeVersionv4. 0"/ >
        </ handlers >
      </ system. webServer >
      < system. net >
        < defaultProxy enabled = "true"/ >
          < settings >
          </ settings >
      </ system. net >
      < dotNetOpenAuth >
        < messaging >
          < untrustedWebRequest >
            < whitelistHosts >
            </ whitelistHosts >
          </ untrustedWebRequest >
        </ messaging >
        < reporting enabled = "true"/ >
        < openid >
          < relyingParty >
            < security requireSsl = "false" >
            </ security >
            < behaviors >
              < add
type = "DotNetOpenAuth. OpenId. RelyingParty. Behaviors. AXFetchAsSregTransform,
```

```
DotNetOpenAuth.OpenId.RelyingParty"/>
        </behaviors>
      </relyingParty>
    </openid>
  </dotNetOpenAuth>
  <uri>
      <idn enabled="All"/>
      <iriParsing enabled="true"/>
  </uri>
  <applicationSettings>
      <psjlmvc4.Properties.Settings>
        <setting name="Projectcode" serializeAs="String">
            <value>2012-0001</value>
        </setting>
        <setting name="Supervisor" serializeAs="String">
            <value>ABCD</value>
        </setting>
        <setting name="CreateDate" serializeAs="String">
            <value>10/07/2012 12:32:00</value>
        </setting>
      </psjlmvc4.Properties.Settings>
  </applicationSettings>
</configuration>
```

在"<connectionStrings>"一节中有如下一子节：

```
<add name="psjldb12Context" connectionString="Data Source=(local);Initial Catalog=psjldb12;Persist Security Info=True;User ID=wgx;Password=wgx;MultipleActiveResultSets=True" providerName="System.Data.SqlClient"/>
```

其作用是模型与数据库的连接字符串，name="psjldb12Context"中的"psjldb12Context"是模型目录Models中类文件"psjldb12Context.cs"的文件名，此文件包含模型类与数据库表的对应关系，是模型类对应的持久性存储对象列表（有关"psjldb12Context.cs"文件将在后续章节中说明）。

本章小结

本章内容从 VS2013 开发工具概述开始，说明了 VS2013 的特点及新建项目的过程和项目结构，在此基础上，完整地介绍了"建设工程监理信息系统"的整体项目目录内容结构以及特殊的系统配置文件 Web.config 的内容结构和本项目所使用的 Web.config 具体内容。

3 ASP.NET MVC 架构及其应用

MVC 是 Web 应用项目开发的一种模式，是一种基于多层结构的 Web 应用系统架构，这种多层结构一方面为数据的远程处理提供更有效的缓冲机制，另一方面为基于 Web 应用系统开发的程序员的开发工作减轻了工作量，也有效地缩短了项目开发周期。MVC 架构将开发工作进行了明确分解，代表着未来 Web 应用开发的一种趋势，因此，微软（Microsoft）公司从 VS2012 开始，已经将 MVC 开发模式内置于其中。本章内容如下：

3.1　ASP.NET MVC 概述
3.2　ASP.NET MVC 项目的运行
3.3　ActionResult 与视图
3.4　Razor 视图引擎

3.1　ASP.NET MVC 概述

模型—视图—控制器（MVC）体系结构模式将应用程序分成三个主要组件：模型、视图和控制器。ASP.NET MVC 框架是一种用于创建 Web 应用程序的 ASP.NET Web 窗体模式的替代模式。ASP.NET MVC 框架是一个可测试性非常高的轻型演示框架，与基于 Web 窗体的应用程序一样，它集成了现有的 ASP.NET 功能，如母版页和基于成员资格的身份验证。MVC 框架在 System.Web.Mvc 程序集中定义。

3.1.1　ASP.NET 简介

ASP.NET 是.NET Framework 的一部分，是微软公司推出的 Web 应用开发技术，是 ASP 技术和.NET Framework 技术相结合的产物。ASP.NET 平台解决了服务端代码嵌入网页中执行的技术问题，也就是服务端脚本语句的前端执行。

3.1.1.1　ASP 和 ASP.NET

ASP 是 Active Server Pages（动态服务器页面），运行于 IIS（Internet Information Server 服务，是 Windows 开发的 Web 服务器）之上的程序，ASP.NET 是在

ASP 的基础上扩展而形成的一种 Web 应用开发技术。

作为战略产品，ASP. NET 提供了一个统一的 Web 开发模型，其中包括开发人员生成企业级 Web 应用程序所需的各种服务。ASP. NET 的语法在很大程度上与 ASP 兼容，同时它还提供一种新的编程模型和结构，可生成伸缩性和稳定性更好的应用程序，并提供更好的安全保护。可以通过在现有 ASP 应用程序中逐渐添加 ASP. NET 功能，即随时增强 ASP 应用程序的功能，是一个可以编译执行的、基于. NET 的环境，可以用任何与. NET 兼容的语言（包括 Visual Basic. NET、C# 和 JScript. NET）创作应用程序。另外，任何 ASP. NET 应用程序都可以使用整个. NET Framework。开发人员可以方便地获得这些技术的优点，其中包括托管的公共语言运行库环境、类型安全、继承等等。

ASP. NET 可以无缝地与 WYSIWYG HTML 编辑器和其他编程工具（包括 Microsoft Visual Studio. NET）一起工作。这不仅使得 Web 开发更加方便，而且还能提供这些工具必须提供的所有优点，包括开发人员可以将服务器控件拖放到 Web 页的 GUI 和完全集成的调试支持。微软为 ASP. NET 设计了一些策略：易于写出结构清晰的代码、代码易于重用和共享、可用编译类语言编写等，目的是让程序员更容易开发出 Web 应用，满足计算向 Web 转移的战略需要。

3.1.1.2 ASP. NET 的新性能

ASP. NET 提供了应用系统更高的稳定性和优秀的升级性，提供项目的快速开发、简便管理、全新语言以及网络服务功能，整个 ASP. NET 的主题就是系统帮用户做了大部分不重要的琐碎的工作，其特点如下：

（1）全新的构造：新的 ASP. NET 引入托管代码（Managed Code）这样一个全新概念，贯穿整个视窗开发平台。托管代码在 NGWS Runtime 下运行，而 NGWS Runtime 是一个时间运行环境，它管理代码的执行，使程序设计更为简便。

（2）高效率：对于一个程序，开发和运行速度是非常令人关注的。一旦代码开始工作，接下来你就得尽可能地让它运作得快些快些再快些。在 ASP 中你只有尽可能精简你的代码，以至于不得不将它们移植到一个仅有很少一点性能的部件中。而现在，ASP. NET 会妥善地解决这一问题。

（3）易控制：在 ASP. NET 里，你将会拥有一个"Data – Bounds"（数据绑定），这意味着它会与数据源连接，并会自动装入数据，使控制工作简单易行。

（4）语言支持：ASP. NET 支持多种语言，包括编译类语言，比如 VB、VC、C#、J#等，在 VS2013 系统中这些都已经内嵌，而且比这些编译类语言独立运行速度快，非常适合编写大型应用项目。

（5）更好的升级能力：快速发展的分布式应用也需要速度更快、模块化程度更强、操作更加简单、平台支持更多和重复利用性更强的设计开发，这需要一种新的技术来适应不同的系统，网络应用和网站运行需要提供更加强大的可升级服务和可扩展性，而 ASP. NET 能够适应上述的各种要求。

3.1.1.3 ASP 与 ASP. NET 的区别

ASP 与 ASP. NET 属于同一系列，但二者还有着本质的区别。

（1）开发语言不同：ASP 仅局限于使用 non - type 脚本语言来开发，在 Web 页中添加 ASP 代码的方法与客户端脚本中添加代码的方法相同，容易导致代码杂乱；ASP. NET 允许程序员选择并使用功能完善的 strongly - type 编程语言，将前端 HTML 代码和后台 ASP 服务端代码分离，并通过功能强大的 . NET Framework 平台，编译运行，提高速度。

（2）运行机制不同：ASP 是解释运行的编程框架，所以执行效率较低；ASP. NET 是编译性的编程框架，运行的是服务器上的编译好的公共语言运行时库代码，可以利用早期绑定，实施编译来提高效率。

（3）开发方式：ASP 把界面设计和程序设计混在一起，维护和重用困难；ASP. NET 把界面设计和程序设计以不同的文件分离开，重用性和维护性得到了提高。

服务端控件的使用及面向对象程序设计是 ASP. NET 技术的最大特点，是典型的 Web Form 开发模式。

3.1.2 MVC 设计模型

MVC 是指"Model"、"View"、"Controller"三个单词的首字母缩写，意为"模型（Model）"、"视图（View）"和"控制器（Controller）"，简称为"MVC"。

MVC 原型来自于 Desktop 程序设计思想，"M"指数据模型，"V"指用户界面，"C"指控制器。其基本思想是：将 M 和 V 的实现代码分离，通过 C 加以控制管理，从而使同一个程序可以使用不同的表现形式。比如一批统计数据你可以分别用柱状图、饼图来表示。C 存在的目的是确保 M 和 V 的同步，一旦 M 改变，V 应该同步更新，反之亦然，从实际的例子可以看出 MVC 就是 Observer 设计模式的一个特例。

3.1.2.1 Web 应用的结构

MVC 是一个设计模式，它强制性地使应用程序的输入、处理和输出分开。使用 MVC，应用程序的结构被分成模型、视图、控制器三个部分，将数据处理、存储、显示分离，各自处理自己的任务。

（1）视图：视图是用户看到并与之交互的界面。对老式的 Web 应用程序来说，视图就是由 HTML 元素组成的界面；在新式的 Web 应用程序中，HTML 依旧在视图中扮演着重要的角色，但一些新的技术已层出不穷，它们包括 Macromedia Flash 和像 XHTML、XML/XSL、WML 等一些标识语言和 Web Services。

如何处理应用程序的界面变得越来越具有挑战性？MVC 一个大的好处是它

能为你的应用程序处理很多不同的视图。在视图中其实没有真正的处理发生，不管这些数据是联机存储的还是一个雇员列表，作为视图来讲，它只是作为一种输出数据并以允许用户操纵的方式。

（2）模型：模型表示企业数据和业务规则。在 MVC 的三个部件中，模型拥有最多的处理任务。例如它可能用 Java、C#等语言和构件对象来处理数据库。被模型返回的数据是中立的，就是说模型与数据格式无关，这样一个模型能为多个视图提供数据。由于应用于模型的代码只需写一次就可以被多个视图重用，所以减少了代码的重复性。

（3）控制器：控制器接受用户的输入并调用模型和视图去完成用户的需求。所以当单击 Web 页面中的超链接和发送 HTML 表单时，控制器本身不输出任何东西也不做任何处理，它只是接收请求并决定调用哪个模型构件去处理请求，然后再确定用哪个视图来显示返回的数据。

3.1.2.2　MVC 模型的特点

首先，最重要的一点是多个视图可以共享一个模型，现在需要用越来越多的方式来访问你的应用程序。对此，其中一个解决之道是使用 MVC，无论你的用户想要 Flash 界面或是 WAP 界面，用一个模型就能处理它们。由于你已经将数据和业务规则从表示层分开，所以你可以最大化地重用你的代码。

由于模型返回的数据没有进行格式化，所以同样的构件能被不同界面使用。例如，很多数据可能用 HTML 来表示，但是它们也有可能要用 Adobe Flash 和 WAP 来表示。模型也有状态管理和数据持久性处理的功能，例如，基于会话的购物车和电子商务过程也能被 Flash 网站或者无线联网的应用程序所重用。

因为模型是自包含的，并且与控制器和视图相分离，所以很容易改变你的应用程序的数据层和业务规则。如果你想把你的数据库从 MySQL 移植到 Oracle 或 SQL Server，或者改变你的基于 RDBMS 数据源到 LDAP，都只需改变你的模型即可。一旦你正确地实现了模型，不管你的数据来自数据库或是 LDAP 服务器，视图将会正确地显示它们。由于运用 MVC 的应用程序的三个部件是相互独立的，改变其中一个不会影响其他两个，所以依据这种设计思想便可以构造良好的松散耦合构件。

控制器也提供了一个好处，就是可以使用控制器来联结不同的模型和视图去完成用户的需求，这样控制器可以为构造应用程序提供强有力的手段。给定一些可重用的模型和视图，控制器可以根据用户的需求选择模型进行处理，然后选择视图将处理结果显示给用户。

3.1.2.3　MVC 的优势和劣势

MVC 的优势主要体现在以下几个方面：

（1）低耦合性。视图层和业务层的分离，允许更改视图层代码而不用重新编译模型和控制器代码；同样，一个应用的业务流程或者业务规则的改变只需要改动 MVC 的模型层即可。因为模型与控制器和视图相分离，所以很容易改变应用程序的数据层和业务规则。

（2）高重用性和可适用性。MVC 模式允许使用各种不同样式的视图来访问同一个服务器端的代码。它包括任何 Web（HTTP）浏览器或者无线浏览器（WAP），比如，用户可以通过电脑也可通过手机来订购某样产品，虽然订购的方式不一样，但处理订购产品的方式是一样的。由于模型返回的数据没有进行格式化，所以同样的构件能被不同的界面使用。例如，很多数据可能用 HTML 来表示，但是也有可能用 WAP 来表示，而这些表示所需要的命令是改变视图层的实现方式，而控制层和模型层无需做任何改变。

（3）较低的生命周期成本。MVC 使开发和维护用户接口的技术含量降低。

（4）快速的部署。使用 MVC 模式使开发时间得到相当大的缩减，它使程序员（Java 或 C#等开发人员）集中精力于业务逻辑，界面程序员（HTML 和 JSP 开发人员）集中精力于表现形式上。

（5）可维护性。分离视图层和业务逻辑层也使得 Web 应用更易于维护和修改。

（6）有利于软件工程化管理。由于不同的层各司其职，每一层不同的应用具有某些相同的特征，有利于通过工程化、工具化管理程序代码。

MVC 的劣势主要表现在以下几个方面：

（1）缺乏明确的结构定义，理论上是分为模型、视图和控制器三个部分，但不同的开发工具对 MVC 结构的设计有所不同，原理也不尽相同，所以完全理解 MVC 并不是很容易，需要花费一些时间去思考。

（2）由于模型和视图要严格的分离，这样也给调试应用程序带来了一定的困难。每个构件在使用之前都需要经过彻底的测试。一旦构件经过了测试，就可以毫无顾忌地重用它们了。

（3）使用 MVC 同时也意味着你将要管理比以前更多的文件，这一点是显而易见的。这样好像我们的工作量增加了，但是请记住这比起它所能带给我们的好处是不值一提的。

MVC 设计模式是一个很好创建软件的途径，它所提倡的一些原则，像内容和显示互相分离可能比较好理解。但是如果你要隔离模型、视图和控制器的构件，你可能需要重新思考你的应用程序，尤其是应用程序的构架方面。如果你肯接受 MVC，并且有能力应付它所带来的额外的工作和复杂性，MVC 将会使你的软件在健壮性、代码重用和结构方面上一个新的台阶。

3.1.3 MVC 运行机制

MVC 结构模型及其运行机制如图 3-1 所示。

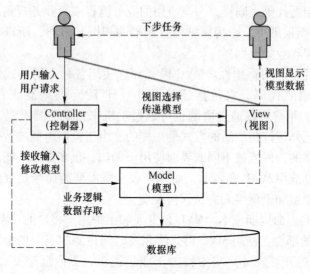

图 3-1 MVC 结构模型

在 MVC 架构下,信息(消息或数据)的具体处理过程为:首先,用户通过浏览器提出请求,调用相应控制器中对应的方法(在控制器类中的动态方法),并传递信息(消息,参数);其次,控制器接收之后,根据要求访问数据模型,修改模型,选择相应的视图并传递模型或消息,或者存储模型数据至数据库。

模型中各部分的作用如图 3-2 所示。

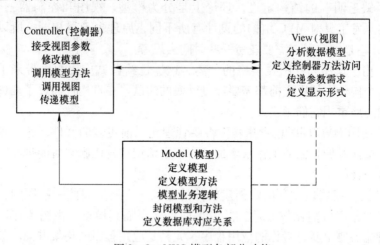

图 3-2 MVC 模型各部分功能

下面说明 MVC 模型各部分之间的关系。

M 和 V:M 和 V 完全分离,不存在直接关系,而是通过 C 进行数据和信息交换。

C 和 M:C 和 M 之间存在调用关系,C 调用 M 中的方法,发起对话的是 C,

而做出回答的是 M，C 可以问 M 各种各样的问题（调用），但 M 只是回答 C 的问题或要求，M 不可以主动向 C 提出要求。所以，C 知道 M 的所有事情，如果用代码来说明这件事情，就是 C 可以导入 M 的头文件或是 M 的接口（API），因为 C 可以通过 M 的 API。

C 和 V：C 接受来自 V 的请求和参数，V 接受 C 传递过来的 M。C 选择相应的 V，来打开并显示 M 中的数据。

M 和 V 没有直接关系，那 M 和 V 之间是否可以交流？不能，否则，控制逻辑会发生混乱，也就是说，如果 M 和 V 存在直接关系，则 C 就会失去意义。

3.1.4 ASP.NET MVC 的特点

ASP.NET MVC 是微软官方提供的以 MVC 模式为基础的 ASP.NET Web 应用程序（Web Application）框架，它由 Castle 的 MonoRail 而来，目前最新版本是 ASP.NET MVC 5.2，并成为 VS2013 内置的 Web 应用开发模板，提供了一种可以代替 ASP.NET WebForm 的基于 MVC 设计模式的应用，是三种 ASP.NET 编程模式中的一种，完全继承了 MVC 的设计模型。

ASP.NET MVC 相对于 ASP.NET WebForm 应用，具有如下优势：

（1）分离任务（输入逻辑、业务逻辑和显示逻辑），易于测试和默认支持测试驱动开发（TDD）。所有 MVC 用到的组件都是基于接口并且可以在进行测试时进行 Mock，在不运行 ASP.NET 进程的情况下进行测试，使得测试更加快速和简捷。

（2）可扩展的简便的框架。MVC 框架被设计用来更轻松的移植和定制功能。你可以自定义视图引擎、UrlRouting 规则及重载 Action 方法等。MVC 也支持 Dependency Injection（DI，依赖注入）和 Inversion of Control（IoC，控制反转）。

（3）强大的 UrlRouting 机制让你更方便地建立容易理解和可搜索的 Url，为 SEO 提供更好的支持。Url 可以不包含任何文件扩展名，并且可以重写 Url 使其对搜索引擎更加友好。

（4）可以使用 ASP.NET 现有的页面标记、用户控件、模板页。你可以使用嵌套模板页，嵌入表达式 <%=%>，声明服务器控件、模板，数据绑定、定位等，自 MVC3 开始，提供了新的 View 视图引擎 Razor，取代了 <%=%> 语法结构。

（5）对现有的 ASP.NET 程序的支持，MVC 让你可以使用如窗体认证和 Windows 认证、Url 认证、组管理和规则、输出、数据缓存、Session、Profile、Health Monitoring、配置管理系统以及 Provider Architecture 特性。

（6）项目资源管理采用"事先约定"的方法建立目录（命名空间）结构，路由清晰，便于程序设计与编写。

3.2 ASP.NET MVC 项目的运行

ASP.NET MVC 项目建成后，其资源目录结构是以"Controllers"、"Views"、"Models"为核心进行配置，项目起始入口（开始首页）是位于"Views"目录下"HomeController"对应的视图目录"Home"中的"Index.cshtml"视图，项目所有视图的访问路由规则是"Controller/Action/id"。

3.2.1 路由规则定义

"路由规则"规定了 MVC 项目中，视图被访问的路径规则，其定义内容位于 App_Start 目录下的"RouteConfig.cs"类文件中。

ASP.NET MVC 应用项目创建之后，系统默认创建了相应的文件夹进行不同层次的开发，"App_Start"是存放项目运行时首先执行的类文件，是 ASP.NET 资源目录之一。

3.2.1.1 RouteConfig.cs 类文件

RouteConfig.cs 的内容如下：

```
using System.Web.Mvc;
using System.Web.Routing;
namespace psjlmvc4
{
    public class RouteConfig
    {
        //注册路由
        public static void RegisterRoutes(RouteCollection routes)
        {
            routes.IgnoreRoute("{resource}.axd/{*pathInfo}");
            //配置路由
            routes.MapRoute(
                name:"Default",
                url:"{controller}/{action}/{id}",
                defaults:new{controller = "Home", action = "Index", id = UrlParameter.Optional}
            );
        }
    }
}
```

在 ASP. NETMVC 应用程序的运行过程中，请求会被发送到 Controller 中，这样就对应了 ASP. NET MVC 应用程序中的 Controllers 文件夹中对应的 Controller；根据路由规则，调用对应的方法（Action），在方法中只负责数据的读取和页面逻辑的处理。

在 Controllers 读取数据时，需要通过 Models 从数据库中读取相应的信息。读取数据完毕后，Controllers 再将数据和 Controller 整合并提交到 Views 视图中，整合后的页面将通过浏览器呈现在用户面前，首次访问时，调用的方法是"Home/Index"，这是由参数"defaults：new ｛controller ="Home"，action ="Index"，id = UrlParameter. Optional｝"定义，是应用程序的入口。

3.2.1.2　其他页面的访问

例如，当用户访问 http：//localhost/Home/About 页面时，首先这个请求会被发送到 Controllers 文件夹中的"HomeController"控制器中对应的方法（Action）"About"，方法"About"中有准备显示数据（模型）的逻辑，然后通过"Return View（）"命令调用位于"Views/Home"文件夹中的"About. cshtml"视图文件，并在浏览器中显示内容，供终端用户使用。下面是控制器"HomeController. cs"中方法（Action）的内容：

```
……
namespace psjlmvc4. Controllers
｛
    [HandleError( )]
    public partial class HomeController：Controller
    ｛
        ……
        //系统门户说明
        [AllowAnonymous]
        public virtual ActionResult About( )
        ｛
            ViewBag. Message ="关于公司情况说明"；
            return View( )；
        ｝
        ……
    ｝
｝
```

View（）没有指定视图名称，则默认的视图是和方法名称相同的视图，其名称是"About. cshtml"，其简化后的内容如下所示：

```
@{
    viewbag.title = "关于我们";
}
<div style = "width:1024px;margin:auto;">
    <div class = "insetText">
        <h2>@ViewBag.Title</h2>
    </div>
    <div style = "border-radius:5px;background-color:#7db9e8;padding:10px;">
        @Html.Label("公司名称:")@psjlmvc4.Properties.Resources.AppliedUnit<br/>
        @Html.Label("公司地址:")@psjlmvc4.Properties.Resources.AppliedUnitAddress<br/>
        @Html.Label("联系电话:")@psjlmvc4.Properties.Resources.AppliedTelephone<br/>
    </div>
</div>
```

运行About.cshtml页面后的实际效果如图3-3所示。

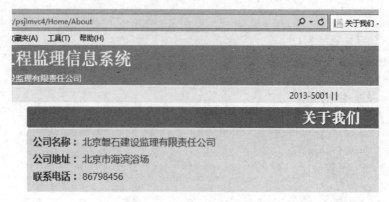

图3-3　About.cshtml页面

3.2.2　路径命名与映射关系

首先控制器类文件的命名规则是"XXXController.cs",例如,"HomeController.cs",其中"XXX"是控制器的标识主名称;第二,在Views目录中,有一个对应的以"XXX"为名称的目录,用于存放"XXXController.cs"中的方法对应的视图文件;第三,"XXXController.cs"的方法"YYY"(关键字ActionResult后的方法名称)的对应视图名称为"YYY.cshtml",存放于"Views/XXX"目录中,如图3-4所示。

ASP.NET MVC应用程序中的URL路径访问是从控制器中的方法开始,并由

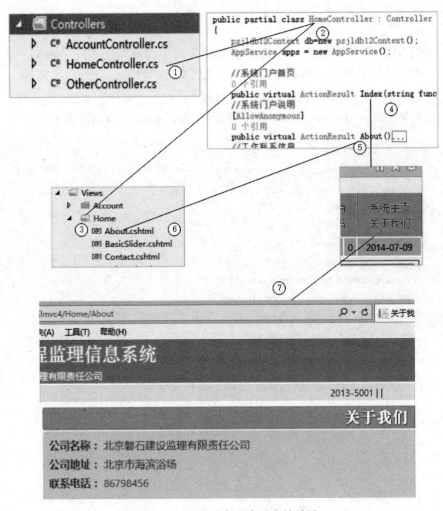

图3-4 MVC 应用程序对应关系图

方法决定调用的页面（视图），例如，访问/Home/About，实际访问的是 HomeControllers 中的 About 方法；而访问/Account/Login 就是访问 AccountControllers 中的 Login 方法，以此类推。

同时，对于目录及文件命名规则，则采用"事先约定"方式，例如，HomeController.cs 控制器类文件对应的视图存放目录是 Views 中的 Home 子文件夹（/Views/Home），而其中的 Index 和 About 方法分别对应于相应目录中的 Index.cshtml 文件和 About.cshtml 文件。

命名约定规则：默认情况下 XXXController.cs 对应 Views 的 XXX 子文件夹，而其中 XXXController.cs 的 YYY() 方法对应 XXX 子文件夹中的 YYY.cshtml，而访问路径 XXX/YYY 即是访问 XXXController.cs 中的 YYY() 方法。

3.2.3 布局页

ASP. NET MVC 架构系统采用了全新的页面整体布局方式——渲染（Render），并通过位于/Views/Shared 目录下的视图文件 "_Layout.cshtml" 实现。

3.2.3.1 布局视图文件_Layout.cshtml

布局视图文件_Layout.cshtml 位于目录/Views/Shared，其内容如下：

```
@{
    var projectcode = Session["projectcode"];
    var projectname = Session["projectname"];
    var logdate = Session["logdate"];
    var userid = Session["userid"];
    var fullname = Session["fullname"];
    var groupid = Session["groupid"];
    var groupname = Session["groupname"];
    var unittypename = Session["unittypename"];
    var unitname = Session["unitname"];
    var funcode = Session["funcode"];
    var mtitle = Session["mtitle"];
}
<!--************************************-->
<!DOCTYPE html>
<html lang="zh">
<head>
    <meta http-equiv="Content-Type" content="text/html;charset=utf-8"/>
    <meta charset="utf-8"/>
    <title>@ViewBag.Title - @psjlmvc4.Properties.Resources.ProjectName</title>
    <link href="~/favicon.ico" rel="shortcut icon" type="image/x-icon"/>
    <meta name="viewport" content="width=device-width"/>
    @Scripts.Render("~/bundles/jquery")
    @Scripts.Render("~/bundles/vsdoc")
    @Scripts.Render("~/bundles/modernizr")
    @Scripts.Render("~/bundles/dateZH")
    @Scripts.Render("~/bundles/jqueryui")
    @Styles.Render("~/Content/themes/base/css")
    @Styles.Render("~/Content/css")
    <script type="text/javascript">
        $(document).ready(function(){
            setInterval('startTime()',1000);
            $(function(){
```

```
                    $(document).tooltip();
                });
            })
            //显示时钟
            function startTime(){
                var today = new Date()
                $('#time-td').text(today.toLocaleTimeString());
            }
            //关闭浏览器    author:wgx
            window.onbeforeunload = function()
            {
                var n = window.event.screenX - window.screenLeft;
                var b = n > document.documentElement.scrollWidth - 20;
                if(b && window.event.clientY < 0 || window.event.altKey)
                {
                    $.get("/psjlmvc4/Home/LogOff");
                }
            }
        </script>
        <style type="text/css" media="print">
            .noprint{display:none;}
        </style>
    </head>
    <body>
        <div class="noprint">
            <table id="table-header">
                <tr id="tr-header">
                    <td id="td-unitlogo">
                        <a href="~/Home/Index">
                            <img id="img-unitlogo" src="~/Images/unitlogo.jpg"/>
                        </a>
                    </td>
                    <td id="td-title">
                        <div class="insetText" style="font-size:2em;">@psjlmvc4.Properties.Resources.ProjectName</div>
                        <div style="font-size:1em;font-family:'Microsoft YaHei';">@psjlmvc4.Properties.Resources.AppliedUnit</div>
                    </td>
                    <td class="td-link" style="border-right:1px solid blue;">
                        @userid - @fullname<br/>
```

```
                    @groupid - @groupname
                  </td>
                  <td class = "td-link"style = "border-right:1px solid blue;">
                    @Html.ActionLink("用户切换","Loging","Home",new{Area=""},null)<br/>
                    @Html.ActionLink("工程选择","pSelect","Home",new{Area=""},null)
                  </td>
                  <td class = "td-link"style = "border-right:1px solid blue;">
@Html.ActionLink("用户注销","LogOff","Home",new{Area="",qexit=true},null)<br/>
                    @Html.ActionLink("修改密码","Updatepwd","Home",new{Area=""},null)
                  </td>
                  <td class = "td-link">
                    @Html.ActionLink("系统主页","Index","Home",new{Area=""},null)<br/>
                    @Html.ActionLink("关于我们","About","Home",new{Area=""},null)
                  </td>
                </tr>
              </table>
              <table id = "table-information">
                <tr>
                  <td id = "td-unittype">@unittypename</td>
                  <td id = "td-unitname">@unitname</td>
                  <td id = "td-project">@projectcode|@projectname|@logdate</td>
                  <td id = "td-scount"
style = "width:20px">@HttpContext.Current.Application["usercount"]</td>
                  <td id = "td-date"
style = "width:100px;">@DateTime.Today.ToString("yyyy-MM-dd")</td>
                </tr>
              </table>
            </div>
            <div id = "div-renderbody">
              @RenderBody()
            </div>
            @RenderSection("scripts",required:false)
          </body>
        </html>
```

如果有其他的多个类似的布局视图文件，就尽量存放于此，因为目录/Views/Shared 是默认优先访问的目录。

3.2.3.2　_Layout.cshtml 内容组成

布局页_Layout.cshtml 是一个独立的网页，是网站中网页的框架，显示内容的模板页，也称母板页，不能直接显示，不能方法直接调用，为统一风格的网站主题做出了贡献。_Layout.cshtml 从内容结构组成可分为三个部分：引入 CSS 和 JS 文件；网站网页的共同显示内容；不同内容网页的渲染（显示）。

第一部分：引入 CSS 和 JS 文件。利用"@Scripts.Render()"引入所需要的 JS 文件；"@Styles.Render()"引入所需要的 CSS 文件。JS 和 CSS 文件可独立引入，也可以由"Bundles.Add()"集合后引入。示例如下：

```
@Scripts.Render("~/Scripts/respond.min.js")//独立文件
@Scripts.Render("~/bundles/jquery")
@Scripts.Render("~/bundles/vsdoc")
@Scripts.Render("~/bundles/modernizr")
@Scripts.Render("~/bundles/dateZH")
@Scripts.Render("~/bundles/jqueryui")
@Styles.Render("~Site.css")//独立文件
@Styles.Render("~/Content/themes/base/css")
@Styles.Render("~/Content/css")
```

"bundles.Add()"方法在位于目录"App_Start"中的类文件"BundleConfig.cs"中完成。"BundleConfig.cs"文件的内容如下：

```
using System.Web;
using System.Web.Optimization;
namespace psjlmvc4
{
    public class BundleConfig
    {
        public static void RegisterBundles(BundleCollection bundles)
        {
            bundles.Add(new ScriptBundle("~/bundles/jquery").Include(
                "~/Scripts/jquery-{version}.js",
                "~/Scripts/bootstrap.min.js",
                "~/Scripts/respond.min.js"));
            bundles.Add(new ScriptBundle("~/bundles/vsdoc").Include(
                "~/Scripts/jquery-{version}-vsdoc.js"));
            bundles.Add(new ScriptBundle("~/bundles/jqueryui").Include(
                "~/Scripts/jquery-ui-{version}.js"));
            bundles.Add(new ScriptBundle("~/bundles/jqueryval").Include(
                "~/Scripts/jquery.validate.js",
                "~/Scripts/jquery.validate.unobtrusive.js"));
```

```
            bundles.Add(new
ScriptBundle("~/bundles/dateZH").Include("~/Scripts/dateZH.js"));
            bundles.Add(new ScriptBundle("~/bundles/modernizr").Include(
                "~/Scripts/modernizr-*"));
            bundles.Add(new StyleBundle("~/Content/css").Include(
                "~/FontAwesome/css/font-awesomr.css",
                "~/Content/bootstrap.min.css",
                "~/Content/psjlmvc4.css",
                "~/Content/WebGrid.css",
                "~/Content/PagedList.css",
                "~/Content/WordEffects.css"));
            bundles.Add(new StyleBundle("~/Content/themes/base/css").Include(
                "~/Content/themes/base/jquery.ui.core.css",
                "~/Content/themes/base/jquery.ui.resizable.css",
                "~/Content/themes/base/jquery.ui.selectable.css",
                "~/Content/themes/base/jquery.ui.accordion.css",
                "~/Content/themes/base/jquery.ui.autocomplete.css",
                "~/Content/themes/base/jquery.ui.button.css",
                "~/Content/themes/base/jquery.ui.dialog.css",
                "~/Content/themes/base/jquery.ui.slider.css",
                "~/Content/themes/base/jquery.ui.tabs.css",
                "~/Content/themes/base/jquery.ui.datepicker.css",
                "~/Content/themes/base/jquery.ui.progressbar.css",
                "~/Content/themes/base/jquery.ui.theme.css",
                "~/Content/themes/base/jquery.ui.tooltip.css"));
        }
    }
}
```

在此引入的 CSS 和 JS 适用于所有以此模板页为母板页的页面使用（不用再次引入）。

第二部分：网站网页的共同显示内容。设计网页共同使用的内容，即模板内容，例如，网站导航、网站说明、页面备注等，例如，_Layout.cshtml 布局页中"<body>"和"</body>"之间的内容。

第三部分：不同内容网页的渲染。注意_Layout.cshtml 布局页中"<body>"和"</body>"之间的内容中，有这样的代码：

```
<div id="div-renderbody">
    @RenderBody()
</div>
```

@RenderBody()是子页渲染并显示的位置（RenderBody()方法），每一个布局页中只能且仅能渲染一次，即"@RenderBody()"只能且仅能出现一次，子页面的内容直接替换到该方法处。

3.2.4 _ViewStart.cshtml 文件

在 ASP.NET MVC 中新建项目之后，Views 目录下自动生成文件"_ViewStart.cshtml"（对应 Razor, C#，也可能是_ViewStart.vbhtml），此文件是在视图被调用之前首先被调用的视图文件，其作用是指定"@RenderBody()"渲染视图所使用的母板页，实现默认 Layout 的作用，内容如下：

```
@{
    Layout = "~/Views/Shared/_Layout.cshtml";
}
```

这个视图文件会在所有视图文件（XXX.cshtml）被执行之前执行，主要用于一些不方便或不能在母版（_Layout.cshtml）中进行的统一操作，例如，你有多个没有继承关系的母版或不使用母版的单页，如果每个视图的母板页执行指向这个文件的操作，虽然没有多大问题，但是重复工作量非常大，Views 目录下_ViewStart.cshtml 文件的使用解决了这个问题。

在_ViewStart.cshtml 文件中，定义一些参数或做一些判断，定义过程和语法与普通的页面定义没有任何差别。_ViewStart 文件可以被用来定义想要在每次视图呈现开始的时候执行的通用视图代码。比如，_ViewStart.cshtml 文件中写下面这样的代码来编程，设置每个视图的默认布局属性为_SiteLayout.cshtml 文件：

```
@{
    Layout = "~/Views/Shared/_SiteLayout.cshtml";
}
```

因为这段代码在每个视图开始的时候执行，所以不需要在任何单个视图文件中显式设置布局（除非我们想要覆盖上面的默认值），实现一次性编写视图逻辑并重复使用，避免在不同的地方重复它。

因此，_ViewStart.cshtml HTML 代码编写的页面文件，可以设计符合用户需要的布局选择逻辑，相对于通过基本属性设置实现的逻辑，控制更加灵活，变化更加多样。例如，可以根据访问网站的设备不同来使用不同的布局模板——有针对手机或 tablet 等这些设备的优化布局，针对 PCs/笔记本的桌面优化布局；或者如果创建一个被不同的用户使用的 CMS 系统或通用共享应用，能根据访问网站的客户（或角色）的不同而选择不同的布局。这将大大提高用户界面的灵活性。也可以在一个控制器或操作筛选器中指定布局页。

3.3 ActionResult 与视图

ASP.NET MVC 项目中创建的控制器（Controller），是 CS 类文件，其内容由不同的方法（函数或 Action）组成，用户请求是根据路径规则发送到控制器中的某个方法（Action），并通过"Return View()"方式返回调用相应视图，显示结果。根据前述的控制器示例，其中大多数方法的返回类型是"ActionResult"，例如，控制器 HomeController 中的 Index() 方法、About() 等，默认的返回类型都是 ActionResult。ActionResult 实际上是一个抽象类（abstract class），位于 System.Web.Mvc 命名空间，而实际返回的类型是该抽象类的子类。

3.3.1 ActionResult 的子类类型

ActionResult 抽象类的子类类型，如表 3-1 所示。

表 3-1 ActionResult 抽象类的子类类型

序号	子类名称	说明
1	ViewResult	表示 HTML 的页面内容
2	EmptyResult	表示空白的页面内容
3	RedirectResult	表示定位到另外一个 URL
4	JsonResult	表示可以运用到 Ajax 程序中 Json 结果
5	JavaScriptResult	表示一个 JavaScript 对象
6	ContentResult	表示一个文本内容
7	FileContentResult	表示一个可以下载的、二进制内容的文件
8	FilePathResult	表示一个可以下载的、指定路径的文件
9	FileStreamResult	表示一个可以下载的、流式的文件

在 MVC 中所有的 Controller 类的内容是由返回类型为"ActionResult"的若干方法（Action）组成，其中对 Action 的要求如下：
(1) 必须是一个 public 方法；
(2) 必须是实例方法；
(3) 没有标志 NonActionAttribute 特性的（NoAction）；
(4) 不能被重载；
(5) 必须返回 ActionResult 类型。
例如，在 HomeController 中，"Index()"方法的定义代码如下：

```csharp
public class HomeController:Controller
{
    //必须返回 ActionResult 类型
    public virtual ActionResult Index(string funcode = null,string mtitle = null)
    {
        ViewBag.Message ="系统门户首页";
        ……
        return View(rd.ToList());
    }
}
```

3.3.2 ActionResult 返回类型说明

3.3.2.1 ActionResult 返回类型：ViewResult

ViewResult 是常用的 ActoinResult 返回类型，对应的返回结果语句是"return View()"，返回 ViewResult 视图结果，将视图呈现给网页。

```csharp
public ActionResult About()
{
    return View();//参数可以返回 model 对象
}
```

3.3.2.2 ActionResult 返回类型：PartialViewResult

返回 PartialViewResult 部分视图结果，主要用于返回部分视图内容（有参数的用户控件）对应的返回结果语句是"return PartialView()"。部分视图文件可根据控制器约定规则，在相应的/Views/目录下创建部分视图，也可以在目录 Views/Shared 下创建（公用且没有参数）。例如，在工程统计功能区域"JArea"的 Views/ProjectStatistics/创建了部分视图"YearCountPartial.cshtml"，对应的调用方法代码如下：

```csharp
//年份工程量统计部分视图方法
public ActionResult YearCountPartial(DateTime date1,DateTime date2)
{
    int y1 = date1.Year,y2 = date2.Year;
    var sl = new StatisticsList(date1,date2);
    var rd = db.AProjectLists.Where(p = >p.logdate.Year > = y1 && p.logdate.Year < = y2);
    for(int i = y1;i < = y2;i + +)
    {
```

```
            sl. xValue. Add( i. ToString( ) ) ;
            sl. yValue. Add( rd. Count( p = > p. logdate. Year = = i ) ) ;
        }
        ViewBag. Message = y1 + " - " + y2 + "年工程量统计";
        return PartialView( sl ) ;
    }
```

页面调用,将"部分视图结果"输出到"部分视图"所在网页相应的位置。

3.3.2.3 ActionResult 返回类型:ContentResult

ContentResult 子类返回用户定义的内容结果,例如:

```
public ActionResult Content( )
{
    return Content( "Test Content", "text/html" ) ;//可以指定文本类型
}
```

则页面输出的内容为"Test Content",此类型多用于在 Ajax 操作中需要返回的文本内容。

3.3.2.4 ActionResult 返回类型:JsonResult

JsonResult 子类返回序列化的 Json 对象,例如:

```
public ActionResult Json( )
{
    Dictionary < string, object > dic = new Dictionary < string, object > ( ) ;
    dic. Add( "id", 100 ) ;
    dic. Add( "name", "hello" ) ;
    return Json( dic, JsonRequestBehavior. AllowGet ) ;
}
```

返回 Json 格式对象"DIC",可以用 Ajax 操作;设置参数"JsonRequestBehavior",否则会提示"此请求已被阻止"错误信息;由于当使用 GET 请求时,会将敏感信息透漏给第三方网站,若要允许 GET 请求,请将 JsonRequestBehavior 设置为"AllowGet"。

3.3.2.5 ActionResult 返回类型:JavaScriptResult

JavaScriptResult 返回可在客户端执行的脚本,例如:

```
public ActionResult JavaScript( )
{
    string str = string. Format("alter('{0}');","返回 JavaScript");
    return JavaScript(str);
}
```

但这里并不会直接响应弹出窗口，需要用页面进行再一次调用。这样可以方便根据不同逻辑执行不同的 js 操作。

3.3.2.6 ActionResult 返回类型：FileResult

FileResult 返回要写入 HTTP 请求响应中的二进制输出文件，一般可以用作需要简单下载的功能，例如：

```
public ActionResult FileDownload( string fullname = null )
{
    var shortname = Path. GetFileName(fullname);
    return File(fullname,"application/x - download", Url. Encode(shortname));
}
```

直接下载 "shortname" 所指定的文件后保存到本地，默认保存文件名为 "shortname"。

3.3.2.7 ActionResult 返回类型：EmptyResult

返回 Null 或者 Void 数据类型的 EmptyResult，不显示内容，例如：

```
public ActionResult Empty( )
{
    return null;
}
```

返回结果为 Null。

3.3.2.8 ActionResult 返回类型：重定向方法类

重定向方法分为 Redirect、RedirectToAction、RedirectToRoute。

（1）Redirect：直接转到指定的 Url 地址。

```
public ActionResult Redirect( )
{
    //直接返回指定的 url 地址
    return Redirect("http://www. baidu. com");
}
```

（2）RedirectToAction：直接使用 Action Name 进行跳转，也可以加上 ControllerName 和参数。

```
public ActionResult RedirectResult()
{
    return RedirectToAction("Index","Home",new{id="100",name="liu"});
}
```

（3）RedirectToRoute：指定路由进行跳转。

```
public ActionResult RedirectRouteResult()
{
    return RedirectToRoute("Default",new{controller="Home",action="Index"});
}
```

Default 为 RouteConfig.cs 中定义的路由名称。

另外需要说明，类"FileContentResult"、"FilePathResult"、"FileStreamResult"是子类"File"的派生类。

3.3.3 View 及其应用

View 即视图，用来在客户端显示用户所请求的结果。视图有布局视图（_Layout.cshtml）、部分视图（PartialView）和方法视图（View），部分视图又分为方法渲染（RenderAction）和直接调用（Partial）两种形式。此处所讨论的视图是方法视图（View）。

3.3.3.1 Controller 与 Action

Controller 是 ASP.NET MVC 中的类，所有新建的控制器继承于此类。控制器的内容是由方法（或者称之为"Action"）组成。下面以新建一个控制器过程为例，说明控制器的应用。

第一步，选中 Controllers 目录，右键弹出菜单中选择"添加（D）"菜单项，再选择"控制器"，即弹出添加控制器选项对话框，如图 3-5 所示。

第二步，确定新建控制器的名称，此处控制器命名为"ProjectListController"，其含义是工程项目管理控制器（与模型类的名称相同），取代默认的名称"Default1Controller"。

第三步，选择控制器模板，决定生成控制器的结构和内容选项，分为以下几个选项：

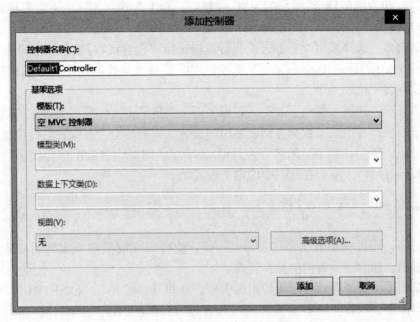

图 3-5　添加控制器选项对话框

（1）空 MVC 控制器：生成只含有 Index（）方法（Action）的控制器，并且不自动生成相应的视图。控制器代码如下：

```
using System. Web. Mvc;
namespace psjlmvc4. Controllers
{
    public class EmptyController:Controller
    {
        public ActionResult Index( )
        {
            return View( );
        }
    }
}
```

可根据需要，增加必要的方法（Action）。

（2）包含读/写操作和视图的 MVC 控制器（使用 Entity Framework）：此选项生成的控制器，会自动生成包括列表显示（Index）、新增（Create）、编辑（Edit）、删除（Delete）操作对应的方法，即 CRUD 操作方法，同时生成对应的视图（根据模板文件生成），当然，"模型类"选项内容不能为空。

(3) 包含空的读/写操作的 MVC 控制器：和上个选项相比，不需要模型类，不生成对应的视图。

第四步，选择模型类，如果控制器模板选择"包含读/写操作和视图的 MVC 控制器（使用 Entity Framework）"选项，则需要指定操作所使用的数据模型类，从下拉列表中选取一个。

第五步，选择数据上下文类。数据上下文类定义了实体模型类对应的数据库存储记录表，并定义了数据存储所使用的表名，根据约定，对应表名为模型类名称加"s"，例如，模型类的名称为"KUserList"，则对应的数据库存储表的名称为"KUserLists"，那么在此类中的定义形式如下：

```
public DbSet < KUserList > KUserLists{get;set;}
```

这样，通过模型数据集"KUserLists"，就可以完成模型类实例的检索、新增、编辑、删除等一系列的操作管理。

第六步，选择生成视图所使用的引擎。这里选择"Razor（CSHTML）"视图引擎。注意，只有控制器模板选项"包含读/写操作和视图的 MVC 控制器（使用 Entity Framework）"有效。

3.3.3.2 建立视图

视图的生成有两种方式：一是在建立控制器时，选择控制器模型"包含读/写操作和视图的 MVC 控制器（使用 Entity Framework）"自动生成；二是建立控制器后选择性生成。

在生成视图时的一个重要选项是"视图引擎"选项，在此选用"Razor"视图引擎。

3.4 Razor 视图引擎

Razor 视图引擎生成的视图中，服务器执行代码设计采用 Razor 语法进行设计编程。Razor 作为一种全新的模板被 MVC 3 及以后版本和 WebMatrix 使用。Razor 在减少代码冗余、增强代码可读性和智能感知方面，都有着突出的优势。

3.4.1 Razor 标识符号

Razor 语句是嵌入 HTML 视图中的服务器端执行代码，同时支持 C#（C sharp）和 VB（Visual Basic），此处只介绍基于 C#（C sharp）语言的 Razor 语法及应用。

3.4.1.1 语法规则

基于 C#语言规则体系的 Razor 引擎语法规则如下：

(1) 代码开始标记符号为"@";
(2) 代码语法规则和C#类似;
(3) 多代码行时需要封装于{…}中,多行时语句行以分号(;)进行分隔;
(4) 变量、函数、Helper等直接引用。

下面是一段嵌入视图（cshtml）中基于Razor引擎的C#代码实例:

```
<!--单行代码块定义变量-->
@{var myMessage = "Hello World";}
<!--行内表达式或变量  使用变量-->
<p>The value of myMessage is:@myMessage</p>
<!--多行语句代码块-->
@{
    var greeting = "Welcome to our site!";
    var weekDay = DateTime.Now.DayOfWeek;
    var greetingMessage = greeting + "Here in Huston it is:" + weekDay;
}
<p>The greeting is:@greetingMessage</p>
@Html.Partial("SubMenuPartial",this.Model)
@Html.ActionLink("A-工程项目","Index",new{funcode = "A",mtitle = "A-工程项目"})
@Html.ActionLink("B-文档管理","Index",new{funcode = "B",mtitle = "B-文档管理"})
@Html.ActionLink("C-前期准备","Index",new{funcode = "C",mtitle = "C-前期准备"})
@Html.ActionLink("D-施工准备","Index",new{funcode = "D",mtitle = "D-施工准备"})
```

3.4.1.2 使用说明

Razor是一种简单的编程语法,类似于<%……%>作用,用于在网页中嵌入服务器端代码。

Razor语法基于ASP.NET框架,该框架是微软的.NET框架特别为Web应用程序开发而设计的组成部分,使用了简化过的语法,网页可被描述为带有两种内容的HTML页面:HTML内容和Razor代码。

当服务器读取这种页面后,在将HTML页面发送到浏览器之前,会首先运行Razor代码。这些在服务器上执行的代码能够完成浏览器中无法完成的任务,比如访问服务器数据库。服务器代码能够在页面被发送到浏览器之前创建动态的HTML内容。从浏览器来看的话,由服务器代码生成的HTML与静态HTML内容没有区别。

3.4.1.3 Razor语法之代码块定义

可以使用@{code}来定义一段代码块,例如:

```
@{
    int num1 = 10;
    int num2 = 5;
    int sum = num1 + num2;
    @ sum;
}
```

在代码块中,我们编写代码的方式和通常服务器端代码的方式是一样的。另外,如果需要输出,例如上面的代码在页面中输出结果,我们可以使用@ sum 完成输出。另外,@(code) 可以输出一个表达式的运算结果,上面的代码也可以写成这样:

```
@{
    int num1 = 10;
    int num2 = 5;
    int sum = num1 + num2;
    @(num1 + num2);
}
```

3.4.1.4 Razor 语法之代码混写

Razor 支持代码混写。在代码块中插入 HTML、在 HTML 中插入 Razor 语句都是可以的,例如:

```
@{
    int num1 = 10;
    int num2 = 5;
    int sum = num1 + num2;
    string color = "Red";
    <font color = "@ color" >@ sum </font>
}
```

3.4.1.5 输出@ 符号

输出 E-mail 地址:Razor 模板会自动识别出 E-mail 地址,所以不需要我们进行任何的转换。而在代码块中,只需要使用@:Tom@ gmail.com 即可。@:表示后面的内容为文本。

输出 HTML 代码(包含标签):直接输出,string html = "< font color = 'red' >文本 </font >";@ html。

输出 HTML 内容（不包含标签）：有两种方法，第一种：IHtmlString html = new HtmlString (″< font color = ′red′ > 文本 </font >″)；@ html；第二种：string html = ″< font color = ′red′ > 文本 </font >″；@ Html. Raw（html）。

3.4.1.6 Razor 语法之注释

这里所说的注释是指服务器端的注释，在 Razor 代码块中，可以使用 C#的注释方式来进行注释，分别是//（单行注释）和/ * */（多行注释）。

另外，Razor 还提供了一种新的服务器短代码注释，可以注释 C#代码，同时可以注释 HTML 代码，@ * * @，这种注释方式不受代码块的限制，在 Razor 代码中的任何位置都可以，例如：

```
@ *
    这是一个注释
    <b> 这个是注释 </b>
* @
```

3.4.2 Razor C#基本语法

变量是用于存储数据的命名实体。

3.4.2.1 变量命名与使用

变量名称必须以字母字符开头，后跟字母或数字或下划线，不能包含空格等其他符号，例如，逗号（,）、句号（.）、惊叹号（!）；不能使用系统所用的保留字，例如，"var"、"int"、"string"。比如：ab12、my_name、projectcode 是合法的；12de、var、my – name、greet ing、create 是非法变量标识符。

变量可以是某个具体的类型，指示其所存储的数据类型。字符串变量存储字符串值，整数变量存储数值，日期变量存储日期值等。

3.4.2.2 变量声明

使用 var 关键词或类型对变量进行声明，例如：

```
var greeting = ″Welcome to W3chtml″;
var counter = 103;
var today = DateTime. Today;
string greeting = ″Welcome to W3chtml″;
int counter = 103;
DateTime today = DateTime. Today;
```

3.4.2.3 数据类型

常用数据类型如下：

(1) int：整数，103，12，5168；
(2) float：浮点数，3.14，3.4e38；
(3) decimal：小数，1037.196543；
(4) bool：逻辑值，true，false；
(5) string：字符串值，"Hello W3chtml"，"Bill"。

3.4.2.4 运算符

Razor 语法引擎支持多种运算符，如表 3-2 所示。

表 3-2 Razor 语法引擎支持的运算符

运算符	描述	实例
=	赋值运算	i=6, a=b
+	加法运算	i=5+5, a=b+c
-	减去值或变量	i=5-5
*	乘值或变量	i=5*5
/	除值或变量	i=5/5
+=	累加运算	S+=5，等价于 S=S+5
-=	累减运算	S-=5，等价于 S=S-5
==	两个等号，等于比较运算	(i==10)，返回 True 或 False
!=	不相等比较运算	(i!=10)，返回 True 或 False
<	小于比较运算	(i<10)，返回 True 或 False
>	大于比较运算	(i>10)，返回 True 或 False
<=	小于等于比较运算	(i<=10)，返回 True 或 False
>=	大于等于比较运算	(i>=10)，返回 True 或 False
+	字符串连接运算	"w3"+"school"="w3school"
.	点运算，分隔对象与方法	DateTime.Hour
()	括号分组运算，优先级最高	(i+5)*6
()	函数运算	x=Add(i, 5)
[ ]	数组运算，访问数组或集合中的值	name[3]
!	逻辑反转运算	!true=false，!false=true
&&	逻辑与运算	true && true = true
||	逻辑或运算	true || true = true

3.4.2.5 数据类型转换与判断函数

Razor 语言引擎中的数据转换与判断函数如表 3-3 所示。

表3-3 Razor 语言引擎中的数据转换与判断函数

方法	描述	实例
AsInt( )	字符串转换为整数	myInt = myString. AsInt ( );
IsInt( )	判断是否整数	if (myString. IsInt ( ))
AsFloat( )	字符串转换为浮点数	myFloat = myString. AsFloat ( );
IsFloat( )	判断是否浮点数	if (myString. IsFloat ( ))
AsDecimal( )	字符串转换为十进制数	myDec = myString. AsDecimal ( );
IsDecimal( )	判断是否十进制数	if (myString. IsDecimal ( ))
IsDateTime( )	判断是否日期时间	if (myString. IsDateTime ( ))
AsDateTime( )	字符串转换为 DateTime 类型	myDate = myString. AsDateTime ( );
AsBool( )	字符串转换为逻辑值	myBool = myString. AsBool ( );
IsBool( )	判断是否逻辑值	if (myString. IsBool ( ))
ToString( )	把任意数据类型转换为字符串	myString = myInt. ToString ( );

3.4.3 Razor C#循环语句

利用循环语句，可以重复执行一组代码。

3.4.3.1 For 循环

在确定循环的次数或循环初值和终值的前提下，可以使用 For 循环，例如，计数或规律步长的循环。

下面是一个 CSHTML 网页中嵌入的显示循环次数的代码。

```
< html >
< body >
@ for( var i = 10 ; i < 21 ; i + + )
   {
       < p > Line@ i < /p >
   }
< /body >
< /html >
```

3.4.3.2 For Each 循环

如果需要处理集合或数组，则通常要用到 For Each 循环。

集合是一组相似的对象，For Each 循环允许在每个项目上执行一次任务。For Each 循环会遍历集合直到完成为止。

例如，遍历 ASP. NET Request. ServerVariables 集合。

```
< html >
< body >
< ul >
@ foreach( var x in Request. ServerVariables)
    { < li > @ x < /li > }
< /ul >
< /body >
< /html >
```

3.4.3.3　While 循环

While 是一种通用的循环。While 循环以关键词 While 开始，后面跟大括号，其中定义循环持续的长度，然后是要循环的代码块。While 循环通常会对用于计数的变量进行增减。

在下面的例子中，循环每运行一次，+ = 运算符就向变量 i 增加 1。

```
< html >
< body >
@ {
    var i = 0;
    while( i < 5 )
    {
        i + = 1;
        < p > Line #@ i < /p >
    }
}
< /body >
< /html >
```

3.4.3.4　数组

数组是用一个变量存储一组类型相同的数据集合。请看下面的示例：

```
@ {
    string[ ] members = {"Jani","Hege","Kai","Jim"} ;
    int i = Array. IndexOf( members ,"Kai") + 1;
    int len = members. Length;
    string x = members[ 2 - 1 ];
}
< html >
```

```
<body>
<h3>Members</h3>
@foreach(var person in members)
{
    <p>@person</p>
}
<p>The number of names in Members are @len</p>
<p>The person at position 2 is @x</p>
<p>Kai is now in position @i</p>
</body>
</html>
```

3.4.4　Razor C#判断语句

3.4.4.1　if（条件）{语句块}

如果条件成立（值为 true），则执行基于条件块内的代码，否则，跳过语句块后执行。如需测试某个条件，使用 if 语句，if 语句基于测试来返回 true 或 false，请看以下示例：

```
@{var price = 50;}
<html>
<body>
@if(price > 30)
{
    <p>The price is too high.</p>
}
</body>
</html>
```

3.4.4.2　if（条件）{语句块1} else {语句块2}

如果条件成立，执行"语句块1"，否则执行"语句块2"，示例如下：

```
@{var price = 20;}
<html>
<body>
@if(price > 30)
{
    <p>The price is too high.</p>
```

```
    else
    {
        <p>The price is OK.</p>
    }
</body>
</html>
```

在上面的例子中，如果价格大于30，则执行"<p>The price is too high.</p>"语句；否则执行"<p>The price is OK.</p>"。

3.4.4.3　if else if

多条件分支，其语法请参考以下示例：

```
@{ var price = 25; }
<html>
<body>
@if( price >= 30 )
{
    <p>The price is high.</p>
}
else if( price > 20 && price < 30 )
{
    <p>The price is OK.</p>
}
else
{
    <p>The price is low.</p>
}
</body>
</html>
```

在上面的例子中，如果第一个条件为true，则执行第一个代码块；否则，如果下一个条件为true，则执行第二个代码块；以此类推；如果所有条件都不成立，则执行"else"后的代码块。

3.4.4.4　switch 条件

switch 是多分支结构，根据条件分支，选择满足条件后的代码块执行，具体示例如下：

```
@ {
    var weekday = DateTime.Now.DayOfWeek;
    var day = weekday.ToString();
    var message = "";
}
< html >
< body >
@ switch( day )
{
    case"Monday":
        message = "This is the first weekday.";
        break;
    case"Thursday":
        message = "Only one day before weekend.";
        break;
    case"Friday":
        message = "Tomorrow is weekend!";
        break;
    default:
        message = "Today is" + day;
        break;
}
< p > @ message < /p >
< /body >
< /html >
```

需要说明：测试值（day）位于括号中。每个具体的测试条件以 case 关键词开头，以冒号结尾，其后允许任意数量的代码行，以 break 语句结尾。如果测试值匹配 case 值，则执行代码行（块）。

switch 代码块可为其余的情况设置默认的 case（default:），允许在所有 case 均不为 true 时执行代码。

3.4.5　几个基于 Razor 帮助器的用法

3.4.5.1　布局（@RenderBody）

ASP.NET MVC 提供 Layout 方式布局，为网站设计统一风格的界面的模板，其"@RenderBody()"方法是模板的核心技术。作为一个母版页，在这个模板框架结构中，"@RenderBody()"出现的位置就是不同网页显示的位置，请看下面的示例：

```
@{
    Layout = "/LayoutPage.cshtml";
    Page.Title = "ASP.NET MVC 布局页";
}
<!DOCTYPE html>
<html lang="en">
    <head>
        <meta charset="utf-8"/>
        <title>我的网站 - @Page.Title</title>
    </head>
    <body>
        <p>This is a layout test</p>
        @RenderBody()
    </body>
</html>
```

3.4.5.2　页面渲染（@RenderPage）

当需要在一个页面中，引用输出另外一个以 Razor 为引擎的页面内容时，比如头部或者尾部这些公共的内容，就可以使用@RenderPage（）方法实现。

例如：在 A 页面中调用输出 B 页面的内容

A 页面的内容如下：

```
<p>
    @RenderPage("/b.cshtml")
</p>
```

B 页面的代码如下：

```
<font color="red">这是一个子页面</font>
```

3.4.5.3　Section 区域渲染（@RenderSection）

Section 是定义在 Layout 中使用的占位符，如果有子页面需要在此位置引用输出，在 Layout 的页面中，使用@RenderSection（"Section 名称"）进行渲染留位，示例如下：

Layout 布局页中的代码：

```
<!DOCTYPE html>
<html lang="en">
```

```
    <head>
        <meta charset="utf-8"/>
        <title>我的网站 - @Page.Title</title>
    </head>
    <body>
        @RenderSection("SubMenu")
            @RenderBody()
    </body>
</html>
```

使用此区域的页面代码:

```
@section SubMenu{
    Hello This is a section implement in About View.
}
```

注意:如果某个子页面不需要在此位置呈现,也就是说没有去实现 SubMenu 功能,则会抛出异常。因此,需要在 Layout 布局页中使用重载 @RenderSection ("SubMenu", false),避免此错误的产生。

```
@if(IsSectionDefined("SubMenu"))
{
    @RenderSection("SubMenu",false)
}
else
{
    <p>SubMenu Section is not defined!</p>
}
```

3.4.5.4 Helper 帮助器渲染 (@helper)

Helper 提供了独立功能代码引用技术,对于可重复使用的方法定义成帮助器,这样不仅可以在同一个页面不同地方使用,还可以在不同的页面使用,例如:

```
@helper sum(int a,int b)
{
    var result = a + b;
```

```
    @result
}//Helper 代码
<div>
    <p>@@helper 的语法</p>
    <p>2+3 = @sum(2,3)</p>
    <p>5+9 = @sum(5,9)</p>
</div>
```

我们通常会把一类 Helper 放在一个单独的 cshtml 文件中，而文件名就相当于一个类名。例如上面的程序段可以改写为如下的形式：

```
@helper sum(int a,int b)
{
    var result = a + b;
    @result
}//Helper 代码
```

把完成 sum 功能的代码放在 HelpMath.cshtml 文件中，则我们在上面的 cshtml 中的使用方法是：

```
<p>2+3 = @HelpMath.sum(2,3)</p>
<p>5+9 = @HelpMath.sum(5,9)</p>
```

另外，系统还为我们提供了一系列的 Helper，用来简化 HTML 的书写。这些 Helper 放在 @Html 中，我们可以方便地使用：

```
<p>
    @Html.TextBox("txtName")
</p>
```

本章小结

本章内容主要介绍了基于 C#语言基础环境下的网页设计架构，以及常用的语言语法规则和技术结构，并较为详细地说明了基于 MVC 模式的项目结构和设计方法。

4 EF 架构与实体模型定义

实体模型（Entity Model）是对数据库中表对象的抽象描述，在 MVC 架构中处于"M"层，被控制层（Controller）处理并传递到视图层（View）进行表现。实体模型定义内容包括数据结构定义、实体关系定义等内容。在 VS2013 系统中设计完成此项任务的工具是 EF 架构，即"实体框架（Entity Framework）"。

本项目中的实体模型定义根据功能模块进行分组归类，即分为工程管理实体模型、文档管理实体模型、前期准备实体模型、施工准备实体模型、进度控制实体模型、施工合同其他事项实体模型、查询统计实体模型、系统管理实体模型、报表实体模型等类别，并存储于相应的区域目录中的"Models"下。为了节省篇幅，本章选取"工程管理"、"文档管理"、"系统管理"功能所涉及的实体模型定义，说明 EF 框架下实体模型定义的过程和方法。

本章主要内容如下：
4.1 EF 概述
4.2 A-工程管理实体模型定义
4.3 B-文档管理实体模型定义
4.4 K-系统管理实体模型定义
4.5 实体模型与数据库的关系

4.1 EF 概述

实体框架（Entity Framework，简称 EF）是一组面向数据模型的管理技术，在 ADO.NET 平台中，支持面向数据管理应用系统的开发。这种数据模型管理技术在 ASP.NET MVC 架构框架下得以充分应用。面向数据的应用程序开发需要实现两个不同的目标：第一是建立实体模型、关系及处理逻辑；第二是处理数据引擎，完成数据处理、数据存储和数据检索。实体框架使开发人员关注于特定域的对象和属性（如客户和客户地址）表单中的数据，无需关心底层的数据库表和存储此数据的位置。在实体框架中，开发人员可以在更高层次上进行抽象工作，并处理数据，相对于传统面向数据的应用程序而言，其创建和维护具有更少的代码。实体框架是.NET 框架的一个组成部分，基于实体框架的应用程序可以运行在已安装.NET Framework 平台的任何计算机系统上（已经集成到 Visual Studio 2010 及以上版本中）。

4.1.1　EF 的特点

现以 ASP.NET Entity Framework 6.1 版本为例，说明其主要特点：

（1）封装性更好，增、删、改、查询实现更加方便。

（2）使用 Linq to Entity 查询，延迟加载（只有在需要时才加载需要的数据）。

（3）开发效率很高，使用 Database First 几乎不用编辑什么代码就能生成一个简单、通用的应用程序。

EF 的工作模式有以下三种：

（1）数据库优先（Database First）：这种方式是比较传统的以数据库为核心的开发模式，比较适合有数据库 DBA 的团队、或者数据库已存在的情况。其优点是编辑代码最少的方式，在有完整的数据库的前提下，你几乎可以不编辑任何代码就能完成应用程序的数据层部分（EF）；其缺点是不够灵活，域模型结构完全由数据库控制生成，结构不一定合理，且受数据库表和字段名影响，命名不规范。

（2）代码优先（Code First）：借助 Code First，可通过使用 C#或 Visual Basic.NET 类来描述模型。模型的基本形状可通过约定来检测。约定是规则集，用于在使用 Code First 时基于类定义自动配置概念模型。

（3）模型优先（Model First）：顾名思义，就是先创建 EF 数据模型，通过数据模型生成数据库的 EF 创建方式。

4.1.2　实体模型（EF）的验证规则

Model 使用 DataAnnotations 定义数据模型并对属性进行验证，首先引入以下的类库：

```
using System.ComponentModel.DataAnnotations;
using System.ComponentModel.DataAnnotations.Schema
```

注意，这里的验证会在 Web 客户端和 EF 端同时进行。验证方法关键字说明如下：

[Key]：数据库：定义类（表）的主键。

[Required]：数据库：会把字段设置成 not null，验证：模型修改时要求必须

输入内容，例如，是否可以为 null，［Required（AllowEmptyStrings = false）］，不能为 null 和空字符串。

［MaxStringLegth］：数据库：字符型字段长度的最大值，模型修改时要求输入内容的长度不能超出此定义的长度。

［MinStringLegth］：最小长度定义，验证模型修改时输入内容的长度是否不够。

［NotMapped］：不和数据库匹配的字段，比如数据库存了 First Name 和 Last Name，我们可以创建一个属性 Full Name，数据库中没有，但是可以使用到其他地方。

［ComplexType］：复杂类型，当你想用一个表，但是表中其他的列做成另外一个类，这个时候可以使用，例如，BlogDetails 是 Blog 表的一部分，在 Blog 类中有个属性是 BlogDetails，如下：

```
[ComplexType]
public class BlogDetails{
    public DateTime? DateCreated{get;set;}
    [MaxLength(250)]
    public string Description{get;set;}
}
```

［ConcurrencyCheck］：这是并发标识，标记为 ConcurrencyCheck 的列，在更新数据前，会检查字段内容有没有改变，如果改变了，说明期间发生过数据修改。这个时候会导致操作失败，出现 DbUpdateConcurrencyException 例外。

［TimeStamp］：这种数据类型表现自动生成的二进制数，确保这些数在数据库中是唯一的。TimeStamp 一般用作给表行加版本戳的机制，存储大小为 8 字节。所以对应的 .NET 类型是 Byte［］，一个表只能有一个 TimeStamp 列。

［Table］［Column］：用来表示数据库匹配的细节，如下例所示：

```
[Table("InternalBlogs")]
public class Blog{
    [Column("BlogDescription",TypeName = "ntext")]
    public String Description{get;set;}
}
```

［DatabaseGenerated］：数据库中有些字段是触发器类型的数据，这些数据不希望在更新的时候使用，但是又想读取出来，可以用 DatabaseGenerated 标记：

```
[DatabaseGenerated(DatabaseGenerationOption.Computed)]
public DateTime DateCreated{get;set;}
```

[ForeignKey]：指定关系所依据的外部键，如下例中的"[ForeignKey("BlogId")]"：

```
public class Post{
public int Id{get;set;}
public string Title{get;set;}
public DateTime DateCreated{get;set;}
public string Content{get;set;}
public int BlogId{get;set;}
[ForeignKey("BlogId")]
public Blog Blog{get;set;}
public ICollection<Comment> Comments{get;set;}}
```

[InverseProperty]：如果子表使用了 ForeignKey，父表又使用子表的 Collection 对象会导致在子表中生成多个外键，这个时候，需要在父表类中的子表 Collection 对象上添加上这个 Attribute，表明会重用子对象的哪个属性作为外键。

其他验证规则将会在实体定义中说明。

4.1.3 EF Code First 默认规则及配置

EF Code First 的默认规则（Convention）包括：

（1）表及列默认规则。EF Code First 默认生成的表名为类名的复数形式，表的生成为 dbo 用户，列名与实体类属性名称相同。

（2）主键约束。实体类中属性名为 Id 或［类名］Id，将作为生成表的主键。若主键为 int 类型，则默认为 Sql Server 的 Identity 类型。

（3）字符类型属性。实体类中 string 类型的属性，在生成表时，对应 Sql Server 中 nvarchar（max）类型。

（4）Byte Array 类型约束。实体类中 byte［］类型的属性，生成表时对应 Sql Server 中 varbinary（max）类型。

（5）Boolean 类型约束。实体类中 bool 类型的属性，在生成表是对应 Sql Server 中 bit 类型。

其关系配置有"一对多"、"多对多"和"一对一"三种，其中"多对多"关系一般情况下需要转换为"一对多"进行处理。

规则配置主要有两种方式：Data Annotations 配置和 Fluent API 配置。

Data Annotations 使用命名空间 System. ComponentModel. DataAnnotations；而 Fluent API 通过重写 DbContext. OnModelCreating 方法实现修改 Code First 默认约束。为保障数据库的稳定性，实际应用中前者较为常用。

4.2 A-工程管理实体模型定义

工程管理模块以字母"A"为代号，区域名称为"Areas/AArea"。主要管理对象有"工程信息"、"工程图片"、"单位工程"、"工程调整"，另外，专设"新增工程"类对象，以方便新增工程功能的实现，没有对应的实体存储表。工程管理实体模型定义存储于名为"AModels. cs"类文件中。

4.2.1 "工程信息"实体模型定义

"工程信息"实体模型内容主要记录工程有关的属性，包括固有属性和因管理所需要而增加的属性，类名为"AProjectList"，对应数据库中存储表的名称为"AProjectLists"，其具体内容如下：

```
//1 ***********************************************
#region 1. AProjectList：工程信息记录模型
[MetadataType(typeof(AProjectList))]
[DisplayName("工程信息")]
public partial class AProjectList:IValidatableObject
{
    public AProjectList()
    {
        this. logdate = DateTime. Today;
        this. adjustdate = DateTime. Today;
        this. begdate = DateTime. Today;
        this. enddate = DateTime. Today;
        this. begpicket = 1;
        this. endpicket = 1;
        this. AProjectImages = new HashSet < AProjectImage > ();
        this. LTaskCheckLists = new HashSet < LTaskCheckList > ();
        this. CAttendDevices = new HashSet < CAttendDevice > ();
        this. CAttendStaffs = new HashSet < CAttendStaff > ();
        this. CAttendUnits = new HashSet < CAttendUnit > ();
        this. CDrawingLists = new HashSet < CDrawingList > ();
        this. CSetupLists = new HashSet < CSetupList > ();
        this. CSupervisionUnits = new HashSet < CSupervisionUnit > ();
```

```csharp
            this.DAttendDesigns = new HashSet<DAttendDesign>();
            this.AAFiles = new HashSet<AAFile>();
            this.ABFiles = new HashSet<ABFile>();
            this.DMeetings = new HashSet<DMeeting>();
            this.DSuperTells = new HashSet<DSuperTell>();
            this.AEfiles = new HashSet<AEFile>();
            this.ACFiles = new HashSet<ACFile>()
            this.HContractLists = new HashSet<HContractList>();
        }
        [Key]
        [Display(Name="工程编号")]
        public string projectcode{get;set;}
        [Display(Name="工程名称")]
        [StringLength(500)]
        public string projectname{get;set;}
        [Display(Name="工程地址")]
        [StringLength(500)]
        public string paddress{get;set;}
        [Display(Name="经手人员")]
        [StringLength(50)]
        public string handman{get;set;}
        [Display(Name="注册日期")]
        [DataType(DataType.Date)]
        [DisplayFormat(DataFormatString="{0:yyyy-MM-dd}",ApplyFormatInEditMode=true)]
        [Required(ErrorMessage="此字段内容为必须填写")]
        public DateTime logdate{get;set;}
        [Display(Name="项目经理")]
        [StringLength(50)]
        public string pmanager{get;set;}
        [Display(Name="开工日期")]
        [DataType(DataType.Date)]
        [DisplayFormat(DataFormatString="{0:yyyy-MM-dd}",ApplyFormatInEditMode=true)]
        public Nullable<System.DateTime> begdate{get;set;}
        [Display(Name="竣工日期")]
        [DataType(DataType.Date)]
        [DisplayFormat(DataFormatString="{0:yyyy-MM-dd}",ApplyFormatInEditMode=true)]
        public Nullable<System.DateTime> enddate{get;set;}
```

```csharp
[Display(Name = "功能说明")]
[StringLength(50)]
public string pfunction{get;set;}
[Display(Name = "质量目标")]
[StringLength(50)]
public string qualitytarget{get;set;}
[Display(Name = "合同价款")]
[DataType(DataType.Currency)]
public Nullable<decimal> contractprice{get;set;}
[Display(Name = "合同模式")]
[StringLength(50)]
public string contractmode{get;set;}
[Display(Name = "规划许可")]
[StringLength(50)]
public string planenable{get;set;}
[Display(Name = "规划证号")]
[StringLength(50)]
public string plancode{get;set;}
[Display(Name = "建设许可")]
[StringLength(50)]
public string buildenable{get;set;}
[Display(Name = "建设证号")]
[StringLength(50)]
public string buildcode{get;set;}
[Display(Name = "开工许可")]
[StringLength(50)]
public string planestablish{get;set;}
[Display(Name = "开工证号")]
[StringLength(50)]
public string establishcode{get;set;}
[Display(Name = "投资性质")]
[StringLength(50)]
public string investproperty{get;set;}
[Display(Name = "投资主体")]
[StringLength(500)]
public string investbody{get;set;}
[Display(Name = "建设面积")]
public Nullable<decimal> builduparea{get;set;}
[Display(Name = "报建面积")]
public Nullable<decimal> reportarea{get;set;}
```

```csharp
[Display(Name = "监理费用", Prompt = "工程监理费用输入")]
[DisplayFormat(ApplyFormatInEditMode = true, DataFormatString = "{0:F}")]
[DataType(DataType.Currency)]
public Nullable<Decimal> supervisionprice { get; set; }
[Display(Name = "调整人员")]
[StringLength(50)]
public string adjustman { get; set; }
[Display(Name = "调整日期")]
[DataType(DataType.Date)]
[DisplayFormat(DataFormatString = "{0:yyyy-MM-dd}", ApplyFormatInEditMode = true)]
public Nullable<System.DateTime> adjustdate { get; set; }
[Display(Name = "开始桩号")]
public Nullable<int> begpicket { get; set; }
[Display(Name = "结束桩号")]
public Nullable<int> endpicket { get; set; }
[Display(Name = "工程级别")]
public string levelid { get; set; }
[Display(Name = "工程性质")]
public string propertyid { get; set; }
[Display(Name = "工程状态")]
public string statusid { get; set; }
[Display(Name = "工程类别")]
public string classid { get; set; }
[Display(Name = "备注说明")]
[StringLength(500)]
public string remark { get; set; }
public IEnumerable<ValidationResult> Validate(ValidationContext vc)
{
    AppService apps = new AppService();
    string dstr1 = apps.GetPropertyDisplayName(typeof(AProjectList), "begdate");
    string dstr2 = apps.GetPropertyDisplayName(typeof(AProjectList), "enddate");
    if(this.begdate > this.enddate)
    {
        yield return new ValidationResult(string.Format("{0}不能在{1}之后！请确认重新输入！", dstr1, dstr2), new[] { "begdate", "enddate" });
    }
    dstr1 = apps.GetPropertyDisplayName(typeof(AProjectList), "begpicket");
    dstr2 = apps.GetPropertyDisplayName(typeof(AProjectList), "begpicket");
    if(this.begpicket > this.endpicket)
```

```
                    {
                        yield return new ValidationResult( string.Format("{0}不能大于{1}！请确认
重新输入!",dstr1,dstr2),new[]{"begpicket","endpicket"});
                    }
                }
        public virtual LStatusList statuslist{get;set;}
        public virtual LClassList classlist{get;set;}
        public virtual LLevelList levellist{get;set;}
        public virtual LPropertyList propertylist{get;set;}
        [ForeignKey("handman")]
        public virtual LUserStaff usHandman{get;set;}
        [ForeignKey("pmanager")]
        public virtual LUserStaff usPmanager{get;set;}
        [ForeignKey("adjustman")]
        public virtual LUserStaff usAdjustman{get;set;}
        public virtual ICollection < AProjectImage > AProjectImages{get;set;}
        public virtual ICollection < AProjectPart > AProjectParts{get;set;}
        public virtual ICollection < LTaskCheckList > LTaskCheckLists{get;set;}
        public virtual ICollection < CAttendDevice > CAttendDevices{get;set;}
        public virtual Collection < CAttendStaff > CAttendStaffs{get;set;}
        public virtual ICollection < CAttendUnit > CAttendUnits{get;set;}
        public virtual ICollection < CSetupList > CSetupLists{get;set;}
        public virtual ICollection < CDrawingList > CDrawingLists{get;set;}
        public virtual ICollection < CSupervisionUnit > CSupervisionUnits{get;set;}
        public virtual ICollection < DAttendDesign > DAttendDesigns{get;set;}
        public virtual ICollection < AAFile > AAFiles{get;set;}
        public virtual ICollection < ABFile > ABFiles{get;set;}
        public virtual ICollection < DMeeting > DMeetings{get;set;}
        public virtual ICollection < DSuperTell > DSuperTells{get;set;}
        public virtual ICollection < AEFile > AEfiles{get;set;}
        public virtual ICollection < ACFile > ACFiles{get;set;}
        public virtual ICollection < HContractList > HContractLists{get;set;}
}
#endregion 1. AProjectList:工程信息记录模型结束
```

"工程信息"实体模型定义内容分别由属性初始化（构造函数中写成）、属性定义、关系定义、内部验证规则四个部分组成。其中，属性有 31 个，主键为"projectcode"。

4.2.2 "工程图片"实体模型定义

"工程图片"实体模型主要记录工程相关图片的属性，类名为"AProjectImage"，对应数据库中存储表的名称为"AProjectImages"，其具体内容如下：

```csharp
//2 ****************************************************
    #region 2. AProjectImage:工程图片记录模型
    [DisplayName("工程图片")]
    [Table("AProjectImages")]
    public class AProjectImage
    {
        [Key]
        [Display(Name ="记录号",ShortName ="序号")]
        public int id{get;set;}
        [Display(Name ="工程编号",ShortName ="工程编号")]
        public string projectcode{get;set;}
        [Display(Name ="文件名称",ShortName ="文件名称")]
        [StringLength(500)]
        public string imagefilename{get;set;}
        [Display(Name ="文件路径",ShortName ="文件路径")]
        [StringLength(500)]
        public string imagefilepath{get;set;}
        [Display(Name ="文件内容",ShortName ="文件内容")]
        public byte[]imagecontent{get;set;}
        [Display(Name ="文件大小",ShortName ="文件大小")]
        public Nullable<int> imagesize{get;set;}
        [Display(Name ="文件类型",ShortName ="文件类型")]
        [StringLength(50)]
        public string imagetype{get;set;}
        [Display(Name ="上传时间",ShortName ="上传时间")]
        [DataType(DataType.Date)]
        [DisplayFormat(DataFormatString ="{0:yyyy-MM-dd}",ApplyFormatInEditMode =true)]
        public Nullable<DateTime> uploaddate{get;set;}
        [Display(Name ="数据库连接",ShortName ="数据库连接")]
        [StringLength(500)]
        public string linkurl{get;set;}
        [Display(Name ="备注说明",ShortName ="备注说明")]
        [StringLength(500)]
        public string remark{get;set;}
        public virtual AProjectList AProjectList{get;set;}
    }
    #endregion 2. AProjectImage:工程图片记录模型
```

4.2.3 "工程增加"实体模型定义

"工程增加"实体模型主要为新增工程管理功能提供临时记录模型,只从"工程信息"模型中抽取 5 个属性组成,类名为"AProjectAdd",在数据库中没有对应存储表,其具体内容如下:

```
//3 ================================================
#region 3. AProjectAdd:工程增加记录模型
[DisplayName("工程增加")]
[NotMapped]
public class AProjectAdd
{
    public AProjectAdd()
    {
        this.logdate = DateTime.Today;
        this.projectcode = String.Empty;
        this.projectname = "修改工程名称";
        this.paddress = "修改工程地址";
    }
    [Key]
    [Remote("CodeValidate","AProject",ErrorMessage = "所输入的工程编号已存在")]
    [Display(Name = "工程编号")]
    [StringLength(50)]
    public string projectcode{get;set;}
    [Display(Name = "工程名称")]
    [StringLength(500)]
    public string projectname{get;set;}
    [Display(Name = "注册日期")]
    [DataType(DataType.Date)]
    [DisplayFormat(DataFormatString = "{0:yyyy-MM-dd}",ApplyFormatInEditMode = true)]
    public DateTime logdate{get;set;}
    [Display(Name = "工程地址")]
    [StringLength(500)]
    public string paddress{get;set;}
    [Display(Name = "经手人员")]
    [StringLength(50)]
    public string handman{get;set;}
}
#endregion 3. AProjectAdd:工程增加记录模型结束
```

4.2.4 "单位工程"实体模型定义

"单位工程"实体模型为工程分部、分解信息提供数据管理记录,其类名为"AProjectPart",在数据库中对应的存储表为"AProjectParts",其具体内容如下:

```
//4 =====================================================
#region 4. AProjectPart:单位工程记录模型
[DisplayName("单位工程")]
[Table("AProjectParts")]
public class AProjectPart
{
    public AProjectPart()
    { }
    [Key]
    [Display(Name = "记录号")]
    [DatabaseGenerated(DatabaseGeneratedOption.Identity)]
    public int AProjectPartID{get;set;}
    [Display(Name = "工程编号")]
    public string projectcode{get;set;}
    [Display(Name = "单位工程名称")]
    [StringLength(500)]
    public string partname{get;set;}
    [Display(Name = "建筑面积(m2)")]
    public Decimal buildingarea{get;set;}
    [Display(Name = "结构类型")]
    [StringLength(50)]
    public string structuretype{get;set;}
    [Display(Name = "地上层数")]
    public int upstorey{get;set;}
    [Display(Name = "地下层数")]
    public int downstorey{get;set;}
    [Display(Name = "总层数")]
    public int totalstorey
    {
        get{return this.upstorey + this.downstorey;}
    }
    [Display(Name = "檐高(m)")]
    public decimal Eaveshight{get;set;}
    [Display(Name = "总高(m)")]
    public decimal allhight{get;set;}
```

```
        [Display(Name="设备安装")]
        [StringLength(50)]
        public string devicesetup{get;set;}
        [Display(Name="工程造价(元)")]
        [DisplayFormat(DataFormatString="{0:c}")]
        [DataType(DataType.Currency)]
        public Decimal buildingcost{get;set;}
        [Display(Name="备注说明")]
        [StringLength(100)]
        public string remark{get;set;}
        public virtual AProjectList AProjectList{get;set;}
}
#endregion 4. AProjectPart:单位工程记录模型结束
```

4.2.5 "工程调整"实体模型定义

"工程调整"实体模型为工程变动信息提供数据管理记录,其类名为"AProjectAdjust",在数据库中对应的存储表为"AProjectAdjusts",其具体内容如下:

```
//5 ================================================
#region 5. AProjectAdjust:工程调整记录模型
[Table("AProjectAdjusts")]
[DisplayName("工程调整")]
public class AProjectAdjust
{
        public AProjectAdjust()
        {
                this.adjustdate = DateTime.Today;
        }
        [Key]
        [Display(Name="记录号")]
        public int id{get;set;}
        [Display(Name="工程编号")]
        [Required()]
        public string projectcode{get;set;}
        [Display(Name="调整内容")]
        [StringLength(500)]
        public string adjustcontent{get;set;}
        [Display(Name="面积调整")]
```

```
                public decimal adjustarea{get;set;}
                [Display(Name="经手人")]
                [StringLength(50)]
                public string handman{get;set;}
                [Display(Name="调整日期")]
                [DataType(DataType.Date)]
[DisplayFormat(DataFormatString="{0:yyyy-MM-dd}",ApplyFormatInEditMode=true)]
                public DateTime adjustdate{get;set;}
                [Display(Name="备注说明")]
                [StringLength(500)]
                public string remark{get;set;}
                public virtual AProjectList AProjectList{get;set;}
        }
        #endregion 5. AProjectAdjust:工程调整记录模型结束
```

4.3　B－文档管理实体模型定义

文档管理模块以字母"B"为代号，文件存储区域名称为"Areas/BArea"。主要管理对象有"接收文件"、"发出文件"、"监理日记（个人登记）"、"监理日志（项目组登记）"，另外，专设"文件类别"类对象，用以记录文件类别信息。文件管理实体模型定义存储于类文件"BModels.cs"中。

4.3.1　"接收文件"实体模型定义

"接收文件"实体模型为接收外部文件信息提供数据管理记录，其类名为"BFileList"，在数据库中对应的存储表为"BFileLists"，其具体内容如下：

```
        //1 ================================================
        #region 1. BFileList:接收文件记录模型
        [DisplayName("接收文件")]
        [Table("BFileLists")]
        public partial class BFileList
        {
                public BFileList()
                {
                        this.createdate = DateTime.Today;
                        this.firstname = "";
```

}
[Key]
[Display(Name="文件编号")]
[StringLength(50)]
[Remote("CodeValidate","BFileList",ErrorMessage="此编号已经存在!")]
[Required()]
public string filecode{get;set;}
[Display(Name="文件名称")]
[StringLength(500)]
[Required()]
public string firstname{get;set;}
[Display(Name="文件副名")]
[StringLength(500)]
public string secondname{get;set;}
[Display(Name="经手人员")]
[StringLength(50)]
public string handname{get;set;}
[Display(Name="收发日期")]
[DataType(DataType.Date)]
[DisplayFormat(DataFormatString="{0:yyyy-MM-dd}",ApplyFormatInEditMode=true)]
public DateTime createdate{get;set;}
[Display(Name="文件类别")]
[StringLength(50)]
public string filetypecode{get;set;}
[Display(Name="发文单位")]
[StringLength(500)]
public string fromtounit{get;set;}
[Display(Name="备注说明")]
[StringLength(500)]
public string remark{get;set;}
public virtual BFileType BFileType{get;set;}
}
#endregion 1. BFileList:接收文件记录模型结束
```

## 4.3.2 "文件类别"实体模型定义

"文件类别"实体模型为收发文件类别信息提供数据管理记录，其类名为"BFileType"，在数据库中对应的存储表为"BFileTypes"，其具体内容如下：
```

```
//2 ================================================
#region 2. BFileType:文件类别记录模型
[DisplayName("文件类别")]
[Table("BFileTypes")]
public class BFileType
{
    public BFileType()
    {
        this.BFileLists = new HashSet<BFileList>();
        this.BSendFiles = new HashSet<BSendFile>();
    }
    [Key]
    [Display(Name="文件类别编号")]
    [StringLength(1)]
    [RegularExpression("^[A-Z]{1}$",ErrorMessage="请输入一位大写字母!")]
    [Remote("CodeValidate","BFileType",ErrorMessage="此编号已经存在!")]
    public string filetypecode{get;set;}
    [Display(Name="文件类别名称")]
    [StringLength(50)]
    public string filetypename{get;set;}
    [Display(Name="文件类别说明")]
    [StringLength(500)]
    public string remark{get;set;}
    public virtual ICollection<BFileList> BFileLists{get;set;}
    public virtual ICollection<BSendFile> BSendFiles{get;set;}
}
#endregion 2. BFileList:文件类别记录模型结束
```

4.3.3 "发出文件"实体模型定义

"发出文件"实体模型为单位所发出的文件信息提供数据管理记录,其类名为"BSendFile",在数据库中对应的存储表为"BSendFiles",其具体内容如下:

```
//3 ================================================
#region 3. BSendFile:发出文件记录模型
[DisplayName("发出文件")]
[Table("BSendFiles")]
public partial class BSendFile
{
```

```csharp
public BSendFile()
{
    this.createdate = DateTime.Today;
    this.firstname = "";
}
[Key]
[Display(Name = "文件编号")]
[StringLength(50)]
[Remote("CodeValidate","BSendFile",ErrorMessage = "此编号已经存在!")]
[Required()]
public string filecode{get;set;}
[Display(Name = "文件名称")]
[StringLength(500)]
[Required()]
public string firstname{get;set;}
[Display(Name = "文件副名")]
[StringLength(500)]
public string secondname{get;set;}
[Display(Name = "经手人员")]
[StringLength(50)]
public string handname{get;set;}
[Display(Name = "收发日期")]
[DataType(DataType.Date)]
[DisplayFormat(DataFormatString = "{0:yyyy-MM-dd}",ApplyFormatInEditMode = true)]
public DateTime createdate{get;set;}
[Display(Name = "文件类别")]
[StringLength(50)]
public string filetypecode{get;set;}
[Display(Name = "接收单位名称")]
[StringLength(500)]
public string fromtounit{get;set;}
[Display(Name = "备注说明")]
[StringLength(500)]
public string remark{get;set;}
public virtual BFileType BFileType{get;set;}
}
#endregion 3. BSendFile:发出文件记录模型结束
```

4.3.4 "监理日记(个人登记)"实体模型定义

"监理日记(个人登记)"实体模型为工程监理管理工作人员提供监理情况信息数据管理记录,其类名为"BDayBook",在数据库中对应的存储表为"BDayBooks",其具体内容如下:

```csharp
//4 ====================================================
#region 4. BDayBook:监理日记(个人登记)记录模型
[DisplayName("监理日记(个人登记)")]
[Table("BDayBooks")]
public partial class BDayBook
{
    public BDayBook()
    {
        this.bookdate = DateTime.Today;
        this.editdate = DateTime.Today;
        this.windlevel = 1;
        this.temperature = 0;
        this.bookname = "修改日记标题...";
        this.bookcontent = "修改日记内容...";
        this.rainlevel = "0";
        this.winddirection = "0";
    }
    [Key]
    [Display(Name = "记录号", Order = 1)]
    public int bookid { get; set; }
    [Display(Name = "工程编号", Order = 2)]
    [StringLength(50)]
    public string projectcode { get; set; }
    [Display(Name = "日记编号", Order = 3)]
    [StringLength(50)]
    public string bookcode { get; set; }
    [Display(Name = "日记标题", Order = 4)]
    [StringLength(50)]
    [Required()]
    public string bookname { get; set; }
    [Display(Name = "日记日期", Order = 5)]
    [DataType(DataType.Date)]
```

```csharp
[DisplayFormat(DataFormatString ="{0:yyyy - MM - dd}",ApplyFormatInEditMode = true)]
public DateTime bookdate{get;set;}
[Display(Name ="天气温度",Order = 6)]
[Range(-60,60,ErrorMessage ="{0}在{1}和{2}之间!")]
public int temperature{get;set;}
[Display(Name ="风向方向",Order = 7)]
[StringLength(50)]
public string winddirection{get;set;}
[Display(Name ="风向级别",Order = 7)]
[Range(1,10,ErrorMessage ="{0}在{1}和{2}之间!")]
public int windlevel{get;set;}
[Display(Name ="雨雪级别",Order = 8)]
[StringLength(50)]
public string rainlevel{get;set;}
[Display(Name ="日记内容",Order = 9)]
[DataType(DataType.MultilineText)]
[Required()]
public string bookcontent{get;set;}
[Display(Name ="记录人",Order = 10)]
[StringLength(50)]
public string bookman{get;set;}
[Display(Name ="修改日期",Order = 11)]
[DataType(DataType.Date)]
[DisplayFormat(DataFormatString ="{0:yyyy - MM - dd}",ApplyFormatInEditMode = true)]
public DateTime editdate{get;set;}
[Display(Name ="备注说明",Order = 12)]
[StringLength(50)]
public string bookclass{get;set;}
[Display(Name ="备注说明",Order = 12)]
[StringLength(500)]
public string remark{get;set;}
public virtual AProjectList AProjectList{get;set;}
// ***********************************************
[NotMapped]//风向方向
public List < SelectListItem > windclass = new List < SelectListItem > ()
{
    new SelectListItem{Value ="A",Text ="东"},
    new SelectListItem{Value ="B",Text ="南"},
```

```
                new SelectListItem{Value = "C", Text = "西"},
                new SelectListItem{Value = "D", Text = "北"},
                new SelectListItem{Value = "E", Text = "东南"},
                new SelectListItem{Value = "F", Text = "东北"},
                new SelectListItem{Value = "H", Text = "西南"},
                new SelectListItem{Value = "I", Text = "西北"},
                new SelectListItem{Value = "O", Text = "无", Selected = true}
            };
        [NotMapped]//雨雪级别
        public List<SelectListItem> rainclass = new List<SelectListItem>()
        {
                new SelectListItem{Value = "A", Text = "大雨"},
                new SelectListItem{Value = "B", Text = "中雨"},
                new SelectListItem{Value = "C", Text = "小雨"},
                new SelectListItem{Value = "D", Text = "大雪"},
                new SelectListItem{Value = "E", Text = "中雪"},
                new SelectListItem{Value = "F", Text = "小雪"},
                new SelectListItem{Value = "H", Text = "晴天"},
                new SelectListItem{Value = "I", Text = "阴天"},
                new SelectListItem{Value = "J", Text = "多云"},
                new SelectListItem{Value = "O", Text = "无", Selected = true}
            };
    }
    #endregion 4. BDayBook：监理日记(个人登记)记录模型结束
```

4.3.5 "监理日志（项目组登记）"实体模型定义

"监理日志（项目组登记）"实体模型为工程监理项目组监理工作情况信息提供数据管理记录，其类名为"BDaySetup"，在数据库中对应的存储表为"BDaySetups"，其具体内容如下：

```
    #region 5. BDaySetup：监理日志(项目组登记)记录模型
        [DisplayName("监理日志(项目组登记)")]
        [Table("BDaySetups")]
    public partial class BDaySetup
    {
        public BDaySetup()
        {
```

```csharp
    this.bookdate = DateTime.Today;
    this.editdate = DateTime.Today;
    this.windlevel = 1;
    this.temperature = 0;
    this.bookname = "修改日记标题...";
    this.bookcontent = "修改日记内容...";
    this.rainlevel = "0";
    this.winddirection = "0";
}
[Key]
[Display(Name = "记录号", Order = 1)]
public int bookid{get;set;}
[Display(Name = "工程编号", Order = 2)]
[StringLength(50)]
public string projectcode{get;set;}
[Display(Name = "日记编号", Order = 3)]
[StringLength(50)]
public string bookcode{get;set;}
[Display(Name = "日记标题", Order = 4)]
[StringLength(50)]
[Required()]
public string bookname{get;set;}
[Display(Name = "日记日期", Order = 5)]
[DataType(DataType.Date)]
[DisplayFormat(DataFormatString = "{0:yyyy-MM-dd}", ApplyFormatInEditMode = true)]
public DateTime bookdate{get;set;}
[Display(Name = "天气温度", Order = 6)]
[Range(-60,60,ErrorMessage = "{0}在{1}和{2}之间!")]
public int temperature{get;set;}
[Display(Name = "风向方向", Order = 7)]
[StringLength(50)]
public string winddirection{get;set;}
[Display(Name = "风向级别", Order = 7)]
[Range(1,10,ErrorMessage = "{0}在{1}和{2}之间!")]
public int windlevel{get;set;}
[Display(Name = "雨雪级别", Order = 8)]
[StringLength(50)]
public string rainlevel{get;set;}
[Display(Name = "日记内容", Order = 9)]
```

```csharp
[DataType(DataType.MultilineText)]
[Required()]
public string bookcontent{get;set;}
[Display(Name ="记录人",Order = 10)]
[StringLength(50)]
public string bookman{get;set;}
[Display(Name ="修改日期",Order = 11)]
[DataType(DataType.Date)]
[DisplayFormat(DataFormatString ="{0:yyyy - MM - dd}",ApplyFormatInEditMode = true)]
public DateTime editdate{get;set;}
[Display(Name ="备注说明",Order = 12)]
[StringLength(500)]
public string remark{get;set;}
public virtual AProjectList AProjectList{get;set;}
// ******************************************
[NotMapped]//风向方向
public List < SelectListItem > windclass = new List < SelectListItem > ()
{
    new SelectListItem{Value ="A",Text ="东"},
    new SelectListItem{Value ="B",Text ="南"},
    new SelectListItem{Value ="C",Text ="西"},
    new SelectListItem{Value ="D",Text ="北"},
    new SelectListItem{Value ="E",Text ="东南"},
    new SelectListItem{Value ="F",Text ="东北"},
    new SelectListItem{Value ="H",Text ="西南"},
    new SelectListItem{Value ="I",Text ="西北"},
    new SelectListItem{Value ="O",Text ="无",Selected = true}
};
[NotMapped]//雨雪级别
public List < SelectListItem > rainclass = new List < SelectListItem > ()
{
    new SelectListItem{Value ="A",Text ="大雨"},
    new SelectListItem{Value ="B",Text ="中雨"},
    new SelectListItem{Value ="C",Text ="小雨"},
    new SelectListItem{Value ="D",Text ="大雪"},
    new SelectListItem{Value ="E",Text ="中雪"},
    new SelectListItem{Value ="F",Text ="小雪"},
    new SelectListItem{Value ="H",Text ="晴天"},
    new SelectListItem{Value ="I",Text ="阴天"},
```

```
                new SelectListItem{Value = "J", Text = "多云"},
                new SelectListItem{Value = "O", Text = "无", Selected = true}
    };
}
#endregion 5. BDaySetup:监理日志(项目组登记)记录模型结束
```

4.4 K – 系统管理实体模型定义

系统管理模块以字母"K"为代号,区域名称为"Areas/KArea"。主要管理对象有:KUserList(系统用户记录模型)、KUserAdd(用户增加模型)、KUserLoging(用户登录记录模型)、KGroupList(系统角色记录模型)、KFunList(系统功能记录模型)、KGroupFun(角色功能记录模型)、KLoginList(用户登录日志记录模型)。系统管理实体模型定义存储于类文件"KModels.cs"中。

4.4.1 "系统用户"实体模型定义

"系统用户"实体模型为本系统的所有用户情况信息提供数据管理记录,其类名为"KUserList",在数据库中对应的存储表对象为"KUserLists",其具体内容如下:

```
#region 1. KUserList:系统用户记录模型
public class KUserList
{
    [Key]
    [Display(Name = "用户标识", ShortName = "用户标识")]
    public string userid{get;set;}
    [Display(Name = "用户口令", ShortName = "用户口令")]
    public string password{get;set;}
    [Display(Name = "用户全名", ShortName = "用户全名")]
    [StringLength(20, ErrorMessage = "{0}的字符在{2}和{1}之间!", MinimumLength =0)]
    public string fullname{get;set;}
    [Display(Name = "用户角色", ShortName = "用户角色")]
    [Required()]
    public string groupid{get;set;}
    [Display(Name = "用户单位", ShortName = "用户单位")]
    public string unitcode{get;set;}
```

```csharp
            [Display(Name="照片目录",ShortName="照片目录")]
            [StringLength(50,ErrorMessage="{0}的长度在{2}和{1}之间。",MinimumLength=0)]
            public string imagepath{get;set;}
            [Display(Name="用户照片",ShortName="用户照片")]
            public byte[] userimage{get;set;}
            [Display(Name="照片类型",ShortName="照片类型")]
            public string mimetype{get;set;}
            [Display(Name="用户说明",ShortName="用户说明")]
            [StringLength(50,ErrorMessage="{0}的长度在{2}和{1}之间。",MinimumLength=0)]
            public string remark{get;set;}
            public virtual KGroupList KGroupList{get;set;}
            public virtual LUnitList LUnitList{get;set;}
        }
        #endregion 1. KUserList:系统用户记录模型结束
```

4.4.2 "用户增加"实体模型定义

"用户增加"实体模型是为系统用户增加管理提供的临时实体模型,其中的属性是从"系统用户"模型中选择的必填项目,其类名为"KUserAdd",在数据库中没有对应的存储表对象。其具体内容如下:

```csharp
        #region 2. KUserAdd:用户增加记录模型
        public class KUserAdd
        {
            [Key]
            [Display(Name="用户标识",Prompt="userid")]
            [Remote("CodeValidate","LUserStaff",ErrorMessage="此用户职员已经存在...")]
            public string userid{get;set;}
            [DataType(DataType.Password)]
            [Display(Name="用户密码",ShortName="用户密码")]
            public string password{get;set;}
            [Display(Name="确认密码",ShortName="确认密码",Description="再次输入密码。")]
            [DataType(DataType.Password)]
            [System.ComponentModel.DataAnnotations.Compare("password")]
            public string confirmpassword{get;set;}
        }
        #endregion 2. KUserAdd:用户增加记录模型结束
```

4.4.3 "用户登录"实体模型定义

"用户登录"实体模型为本系统已注册用户登录系统时的用户情况信息提供数据管理记录，其类名为"KUserLoging"，为临时模型对象，在数据库中没有对应的存储表对象，其具体内容如下：

```csharp
#region 3. KUserLoging:用户登录记录模型
public class KUserLoging
{
    public KUserLoging()
    {
        Random rdm = new Random();
        this.randcode = "";
        for(var i = 0; i < 5; i++)
        {
            this.randcode += rdm.Next(10);
        }
        randcode = randcode.Trim();
    }
    [Display(Name = "用户标识", ShortName = "用户标识")]
    [Required(ErrorMessage = "*")]
    public string userid { get; set; }
    [DataType(DataType.Password)]
    [Display(Name = "用户密码", ShortName = "用户密码")]
    public string password { get; set; }
    [Display(Name = "记住我??", ShortName = "记住我??")]
    public bool rememberme { get; set; }
    [Display(Name = "输入验证码")]
    [System.ComponentModel.DataAnnotations.Compare("randcode", ErrorMessage = "*")]
    public string randinput { get; set; }
    [Display(Name = "随机验证码", ShortName = "随机验证码")]
    public string randcode { get; set; }
}
#endregion 3. KUserLoging:用户登录记录模型结束
```

4.4.4 "系统角色"实体模型定义

"系统角色"实体模型为本系统的角色情况信息提供数据管理记录，以实现功能分组管理功能。其类名为"KGroupList",在数据库中对应的存储表对象为"KGroupLists",其具体内容如下:

```
#region 4. KGroupList:系统角色
[MetadataType(typeof(KGroupList))]//grouplist 就是下面定义的字符规则
public class KGroupList
{
    public KGroupList()
    {
        this.LUserStaffs = new HashSet<LUserStaff>();
    }
    [Key]
    [Display(Name="角色序号")]
    [Required()]
    [Remote("CodeValidate","KGroupList",ErrorMessage="此用户角色已经存在")]
    [RegularExpression("^[A-Z]{1}$",ErrorMessage="{0}必须是一位大写字母")]
    public string groupid{get;set;}
    [Display(Name="角色名称")]
    [StringLength(50,ErrorMessage="{0}的长度不能超过{1}(最多20个汉字)!")]
    public string groupname{get;set;}
    [Display(Name="备注说明")]
    [StringLength(500,ErrorMessage="{0}的长度不能超过{1}(最多50个汉字)!")]
    public string remark{get;set;}
    public virtual ICollection<LUserStaff> LUserStaffs{get;set;}
}
#endregion 4. KGroupList:系统角色结束
```

4.4.5 "系统功能"实体模型定义

"系统功能"实体模型为本系统所有实现的功能情况信息提供数据管理记录,其类名为"KFunList",在数据库中对应的存储表对象为"KFunLists",其具体内容如下:

```
#region 5. KFunList:系统功能记录模型
public class KFunList
{
```

```
public KFunList( )
{
    this. KGroupFuns = new HashSet < KGroupFun > ( );
}
[Key]
[Display(Name = "功能编号")]
[Required]
[StringLength(4)]
[Remote("CodeValidate","KFunList",ErrorMessage = "此编号已存在!")]
public string funcode{get;set;}
[Display(Name = "功能名称")]
[Required( )]
[StringLength(50)]
public string funname{get;set;}
[Display(Name = "方法名称")]
[StringLength(50)]
public string actionname{get;set;}
[Display(Name = "控制器名")]
[StringLength(50)]
public string controllername{get;set;}
[Display(Name = "网页名称")]
[StringLength(50)]
public string aspxpage{get;set;}
[Display(Name = "XAML 名称")]
[StringLength(50)]
public string xamlpage{get;set;}
[Display(Name = "功能选择")]
public Nullable < bool > yesno{get;set;}
[Display(Name = "备注说明")]
[StringLength(100)]
public string remark{get;set;}
public virtual ICollection < KGroupFun > KGroupFuns{get;set;}
}
#endregion 5. KFunList:系统功能记录模型结束
```

4.4.6 "角色功能"实体模型定义

"角色功能"实体模型为本系统的角色所拥有的功能情况信息提供数据管理记录,其数据来自"系统角色"和"系统功能"模型,其类名为"KGroupFun",

在数据库中对应的存储表对象为"KGroupFuns",其具体内容如下:

```
#region 6. KGroupFun:角色功能
public class KGroupFun
{
    [Key]
    [Display(Name="标识号")]
    public int id{get;set;}
    [Display(Name="角色编号")]
    public string groupid{get;set;}
    [Display(Name="功能编号")]
    public string funcode{get;set;}
    public virtual KGroupList KGroupList{get;set;}
    public virtual KFunList KFunList{get;set;}
}
#endregion 6. KGroupFun:角色功能结束
```

需要说明的是,"系统角色"和"系统功能"和此对象是一对多关系。

4.4.7 "用户登录日志"实体模型定义

"用户登录日志"实体模型为本系统的所有用户登录系统并完成操作情况信息提供数据管理记录,其类名为"KLoginList",在数据库中对应的存储表对象为"KLoginLists",其具体内容如下:

```
#region 7. KLoginList:用户登录日志记录模型
public class KLoginList
{
    [Key]
    [Display(Name="序号")]
    public int loginid{get;set;}
    [Display(Name="用户标识")]
    [StringLength(20)]
    public string userid{get;set;}
    [Display(Name="系统角色")]
    [StringLength(20)]
    public string event1{get;set;}
    [Display(Name="行为类型")]
    [StringLength(20)]
    public string actiontype{get;set;}
```

```
        [Display(Name = "登录时间")]
        public Nullable < System.DateTime > logintime{get;set;}
        [Display(Name = "注销时间")]
        public Nullable < System.DateTime > logofftime{get;set;}
        [Display(Name = "计算机名称")]
        public string hostname{get;set;}
        [Display(Name = "IP 地址")]
        public string hostip{get;set;}
    }
    #endregion 7. KLoginList:用户登录日志记录模型结束
```

注：为节省篇幅，其余实体模型定义在此省略。

4.5 实体模型与数据库的关系

实体模型只是数据记录结构定义，实际并不持久保存数据，利用模型处理数据时，需要从数据库中读出相应的数据或数据集合，因此实体数据模型需要和数据库建立联系，以完成数据的读取和存储。在 MVC 框架中，建立模型和数据库的联系是通过继承"DbContext"类并定义模型集合实现的。

4.5.1 模型与 DbContext 类

模型定义之后，需要通过继承 DbContext 类（数据库上下文类）建立模型与数据库表对象之间的集合关系，并在 Web.config 配置文件中定义连接对应的连接字符，实现关联，其关系是：

（1）每一个实体模型在基于 DbContext 类的实例类文件中对应一个数据集合（DbSet）；

（2）实例类文件名是连接数据库的关键字，并以此名称建立与数据库的实际连接字符串，定义在 Web.config 配置文件中。

实体模型、数据库、DbContext 类文件、Web.config 配置文件之间的关系如图 4-1 所示。

图中"AProjectList"是实体模型名称；"psjldb12Context"继承 DbContext 类的上下文类；"psjldb12"是数据库名称；"Web.config"是项目配置文件。

4.5.2 psjldb12Context.cs 类文件

psjldb12Context.cs 类文件定义了项目中所有实体模型对应的数据记录集合，其内容如下：

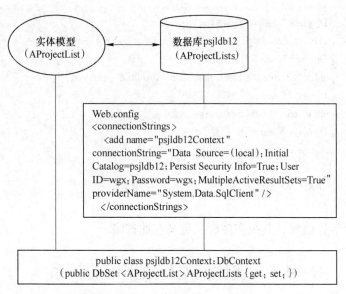

图 4-1 模型与 DbContext 类的关系

```
using System.Data.Entity;
namespace psjlmvc4.Models
{
    public class psjldb12Context:DbContext
    {
        static psjldb12Context()
        {
            Database.SetInitializer < psjldb12Context > (null);
        }
        public psjldb12Context()
            :base("Name = psjldb12Context")
        {
        }
        //监理工作任务及查阅记录
        public DbSet < LTaskItemList > LTaskItemLists{get;set;}
        public DbSet < LTaskCheckList > LTaskCheckLists{get;set;}
        //A - 工程管理
        public DbSet < AProjectList > AProjectLists{get;set;}
        public DbSet < AProjectPart > AProjectParts{get;set;}
        public DbSet < AProjectAdjust > AProjectAdjusts{get;set;}
        //B - 文档管理
        public DbSet < BFileList > BFileLists{get;set;}
```

```csharp
public DbSet < BFileType > BFileTypes { get; set; }
public DbSet < BDayBook > BDayBooks { get; set; }
public DbSet < BSendFile > BSendFiles { get; set; }
public DbSet < BDaySetup > BDaySetups { get; set; }
//C - 前期准备
public DbSet < CAttendDevice > CAttendDevices { get; set; }
public DbSet < CAttendStaff > CAttendStaffs { get; set; }
public DbSet < CAttendUnit > CAttendUnits { get; set; }
public DbSet < CSetupList > CSetupLists { get; set; }
public DbSet < CDrawingCheck > CDrawingChecks { get; set; }
public DbSet < CDrawingCheckItem > CDrawingCheckItems { get; set; }
public DbSet < CDrawingList > CDrawingLists { get; set; }
public DbSet < CContractCheck > CContractChecks { get; set; }
public DbSet < CSupervisionUnit > CSupervisionUnits { get; set; }
//D - 施工准备
public DbSet < DAttendDesign > DAttendDesigns { get; set; }
public DbSet < DMeeting > DMeetings { get; set; }
public DbSet < DSuperTell > DSuperTells { get; set; }
//E - 进度控制
public DbSet < ACFile > ACFiles { get; set; }
public DbSet < AIFile > AIFiles { get; set; }
public DbSet < AIFileStaff > AIFileStaffs { get; set; }
public DbSet < AIFileMaterial > AIFileMaterials { get; set; }
public DbSet < AIFileMachine > AIFileMachines { get; set; }
public DbSet < AOFile > AOFiles { get; set; }
//F - 质量控制
public DbSet < ADFile > ADFiles { get; set; }
public DbSet < ADFileMaterial > ADFileMaterials { get; set; }
public DbSet < ADFileAppend > ADFileAppends { get; set; }
public DbSet < AFFile > AFFiles { get; set; }
public DbSet < AFFileAppend > AFFileAppends { get; set; }
public DbSet < AFFileSubPackage > AFFileSubPackages { get; set; }
public DbSet < AGFile > AGFiles { get; set; }
public DbSet < AGFileAppend > AGFileAppends { get; set; }
public DbSet < AHFile > AHFiles { get; set; }
public DbSet < AHFileAppend > AHFileAppends { get; set; }
//G - 造价控制
public DbSet < AKFile > AKFiles { get; set; }
public DbSet < AKFileAppend > AKFileAppends { get; set; }
public DbSet < ALFile > ALFiles { get; set; }
```

```csharp
public DbSet<ALFileAppend> ALFileAppends{get;set;}
//H-施工合同其他事项
public DbSet<HContractType> HContractTypes{get;set;}
public DbSet<HContractList> HContractLists{get;set;}
public DbSet<HChangeItem> HChangeItems{get;set;}
//I-规程法规
//J-查询统计
//K-系统管理
public DbSet<KFunList> KFunLists{get;set;}
public DbSet<KGroupFun> KGroupFuns{get;set;}
public DbSet<KGroupList> KGroupLists{get;set;}
public DbSet<KUserList> KUserLists{get;set;}
public DbSet<KLoginList> KLoginLists{get;set;}
//L-基础数据
public DbSet<LClassList> LClassLists{get;set;}
public DbSet<LDeviceList> LDeviceLists{get;set;}
public DbSet<LLevelList> LLevelLists{get;set;}
public DbSet<LPropertyList> LPropertyLists{get;set;}
public DbSet<LStaffList> LStaffLists{get;set;}
public DbSet<LStatusList> LStatusLists{get;set;}
public DbSet<LUnitList> LUnitLists{get;set;}
public DbSet<LDepartmentList> LDepartmentLists{get;set;}
public DbSet<LSupervisionUnit> LSupervisionUnits{get;set;}
public DbSet<LUnitType> LUnitTypes{get;set;}
public DbSet<LUserStaff> LUserStaffs{get;set;}
public DbSet<LReportList> LReportLists{get;set;}
public DbSet<LArticleType> LArticleTypes{get;set;}
public DbSet<LArticleList> LArticleLists{get;set;}
//M-安全管理
//O-其他辅助信息管理
public DbSet<OStudyList> OStudyLists{get;set;}
public DbSet<OnlineUser> OnlineUsers{get;set;}
//附录A 连接
public DbSet<AAFile> AAFiles{get;set;}
public DbSet<AAFileList> AAFileLists{get;set;}
public DbSet<ABFile> ABFiles{get;set;}
public DbSet<ACFileAppend> ACFileAppends{get;set;}
public DbSet<AEFile> AEFiles{get;set;}
public DbSet<AEFileList> AEFileLists{get;set;}
public DbSet<AJFile> AJFiles{get;set;}
```

```
        public DbSet < AMFile > AMFiles {get; set;}
        public DbSet < ANFile > ANFiles {get; set;}
        public DbSet < APFile > APFiles {get; set;}
    //附录 B 连接
        public DbSet < BAFile > BAFiles {get; set;}
        public DbSet < BBFile > BBFiles {get; set;}
        public DbSet < BCFile > BCFiles {get; set;}
        public DbSet < BDFile > BDFiles {get; set;}
        public DbSet < BEFile > BEFiles {get; set;}
        public DbSet < BFFile > BFFiles {get; set;}
        public DbSet < BGFile > BGFiles {get; set;}
        public DbSet < BHFile > BHFiles {get; set;}
    //附录 C 连接
        public DbSet < CAFile > CAFiles {get; set;}
        public DbSet < CBFile > CBFiles {get; set;}
    //附录 D 连接
        public DbSet < DAFile > DAFiles {get; set;}
        public DbSet < DBFile > DBFiles {get; set;}
    //附录 E 连接
        public DbSet < EAFile > EAFiles {get; set;}
        public DbSet < EBFile > EBFiles {get; set;}
        public DbSet < EBFileAppend > EBFileAppends {get; set;}
        public DbSet < ECFile > ECFiles {get; set;}
        public DbSet < DTIFile > DTIFiles {get; set;}
        public DbSet < DTHFile > DHTFiles {get; set;}
        public DbSet < OBlogType > OBlogTypes {get; set;}
        public DbSet < OBlogList > OBlogLists {get; set;}
        }
    }
```

对于"public DbSet < ECFile > ECFiles {get; set;}"定义语句,其中"ECFiles"就是实体模型"ECFiles"的数据记录集合,对应于数据库上的表对象"ECFiles"。

4.5.3 Web.config 文件与 < connectionStrings > 节

Web.config 是项目配置文件,定义有关系统运行所需要的环境及其相关参数,其内容如下所示:

```xml
<?xml version="1.0" encoding="utf-8"?>
<configuration>
    <configSections>
        <section name="entityFramework" type="System.Data.Entity.Internal.ConfigFile.EntityFrameworkSection,EntityFramework,Version=6.0.0.0,Culture=neutral,PublicKeyToken=b77a5c561934e089" requirePermission="false"/>
        <sectionGroup name="applicationSettings" type="System.Configuration.ApplicationSettingsGroup,System,Version=4.0.0.0,Culture=neutral,PublicKeyToken=b77a5c561934e089">
            <section name="psjlmvc4.Properties.Settings" type="System.Configuration.ClientSettingsSection,System,Version=4.0.0.0,Culture=neutral,PublicKeyToken=b77a5c561934e089" requirePermission="false"/>
        </sectionGroup>
        <!-- For more information on Entity Framework configuration, visit http://go.microsoft.com/fwlink/?LinkID=237468 -->
        <sectionGroup name="dotNetOpenAuth" type="DotNetOpenAuth.Configuration.DotNetOpenAuthSection,DotNetOpenAuth.Core">
            <section name="oauth" type="DotNetOpenAuth.Configuration.OAuthElement,DotNetOpenAuth.OAuth" requirePermission="false" allowLocation="true"/>
            <section name="openid" type="DotNetOpenAuth.Configuration.OpenIdElement,DotNetOpenAuth.OpenId" requirePermission="false" allowLocation="true"/>
            <section name="messaging" type="DotNetOpenAuth.Configuration.MessagingElement,DotNetOpenAuth.Core" requirePermission="false" allowLocation="true"/>
            <section name="reporting" type="DotNetOpenAuth.Configuration.ReportingElement,DotNetOpenAuth.Core" requirePermission="false" allowLocation="true"/>
        </sectionGroup>
    </configSections>
    <connectionStrings>
        <add name="psjldb12Context" connectionString="Data Source=(local);Initial Catalog=psjldb12;Persist Security Info=True;User ID=wgx;Password=wgx;MultipleActiveResultSets=True" providerName="System.Data.SqlClient"/>
    </connectionStrings>
    <appSettings>
        <add key="webpages:Version" value="2.0.0.0"/>
        <add key="webpages:Enabled" value="false"/>
        <add key="PreserveLoginUrl" value="true"/>
        <add key="ClientValidationEnabled" value="true"/>
        <add key="UnobtrusiveJavaScriptEnabled" value="true"/>
    </appSettings>
    <system.web>
```

```
        < compilation targetFramework = "4.5.1"debug = "true"/ >
        < httpRuntime targetFramework = "4.5.1"/ >
        < authentication mode = "Forms" >
            < forms loginUrl = " ~ /Home/Loging"timeout = "100"protection = "All" >
                < credentials passwordFormat = "SHA1"/ >
            </forms >
        </authentication >
        ..................................................
    </configuration >
```

其中节"< connectionStrings >...</connectionStrings >"是定义数据库连接字符串的节，其内容是：

```
< connectionStrings >
    < add name = "psjldb12Context"connectionString = "Data Source = (local);Initial Catalog = psjldb12;Persist Security Info = True;User ID = wgx;Password = wgx;MultipleActiveResultSets = True"providerName = "System.Data.SqlClient"/ >
</connectionStrings >
```

其中"（local）"是指"localhost"，即本地服务器。

本章小结

本章首先介绍了实体框架的基本概念和定义方法，在此基础上选取了项目中部分实体并说明其定义方法及过程，也包括实体模型与数据库通讯的方法。

5 功能导航系统设计

功能导航系统是信息系统的运行地图,是指导用户完成系统功能的入口。根据权限职责管理的要求,不同类型的用户,其权限职责不同,所分配的功能也不同,因此,功能导航系统的设计与实现,既要考虑数据操作安全,同时需要根据用户的角色进行功能的动态分配。本章内容主要包括:
5.1 系统功能管理
5.2 系统角色管理
5.3 用户角色分配
5.4 角色功能分配
5.5 用户登录与动态功能导航实现

5.1 系统功能管理

建设工程监理信息系统的功能是根据监理任务的先后顺序进行设计,分别由工程管理、文档管理、前期准备、施工准备等 20 个任务模块组成,每个功能模块又由不同的子功能模块组成。系统功能的体系结构为二级体系结构,其内容涵盖了监理工作的各个方面。

实现系统功能管理的控制器和视图等文件位于功能区域 KArea,实现系统功能管理的方法定义在控制器 KFunListController 类中。

5.1.1 功能模块与子功能模块数据记录

系统一级功能模块共设计有 20 个,每个功能模块下又设计若干个子功能,记录系统功能信息的数据模型是 KFunList 类。

一级功能模块的编号使用一位大写英文字母表示,例如"工程管理"功能模块的编号为"A","文档管理"的功能模块编号为"B",以此类推。其相应子功能的编号使用其一级功能模块的编号后跟两位数字组成,例如,"工程管理"共有六个子功能模块,其编号分别"A01"、"A03"、"A04"、"A07"、"A09"、"A10"。

一级功能模块的名称与编号对应关系如表 5-1 所示。

表 5-1 一级功能模块编号与名称对照表

功能编号	功能名称	功能编号	功能名称
A	工程管理	B	文档管理
C	前期准备	D	施工准备
E	进度控制	F	质量控制
G	造价控制	H	施工合同其他事项
I	规程法规	J	查询统计
K	系统管理	L	基础数据
M	安全管理	O	其他辅助信息管理
P	施工单位	Q	建设单位
R	设计单位	S	勘察单位
T	新闻公告管理	U	系统待设功能

其中，P、Q、R、S、U 是为相应的工程建设参与单位的管理工作预留的功能接口。

子功能模块除需要上述数据外，还需要提供其对应功能的入口方法名称（ActionName）及实现方法的控制器名称（ControllerName），即功能操作实现的 URL，以实现功能链接，进入操作界面。对于子功能项目的数据请参阅系统运行时的数据。

5.1.2 系统功能管理控制器

系统功能管理控制器的类名为"KFunListController"，继承于系统"Controller"类，相应的类文件名称为"KFunListController.cs"，其代码内容如下：

```
using psjlmvc4.Models;
using System.Data.Entity;
using System.Linq;
using System.Web.Mvc;
namespace psjlmvc4.Areas.KArea.Controllers
{
    public partial class KFunListController:Controller
    {
        private psjldb12Context db = new psjldb12Context();
        private AppService apps = new AppService();
        [HttpGet]
```

```csharp
public ActionResult CodeValidate(string funcode)
{
    bool exists = true;
    var rd = db.KFunLists.Find(funcode);
    if(rd == null) exists = false;
    return Json(!exists, JsonRequestBehavior.AllowGet);
}
public virtual ActionResult Index(string fcode = null)
{
    string code1 = "";
    if(fcode == null)
    {
        code1 = apps.GetSession("fcode");
    }
    else
    {
        code1 = fcode;
        apps.SetSession("fcode", code1);
    }
    ViewBag.fcode = code1;
    return View(db.KFunLists.Where(f => f.funcode.Contains(code1)).ToList());
}
public virtual ActionResult Details(string id = null)
{
    KFunList funlist = db.KFunLists.Find(id);
    if(funlist == null)
    {
        return HttpNotFound();
    }
    return View(funlist);
}
public virtual ActionResult Create()
{
    return View();
}
[HttpPost]
public virtual ActionResult Create(KFunList funlist)
{
    if(ModelState.IsValid)
    {
```

```csharp
            db.KFunLists.Add(funlist);
            db.SaveChanges();
            return RedirectToAction("Index");
        }
        return View(funlist);
    }
    public virtual ActionResult Edit(string id = null)
    {
        KFunList funlist = db.KFunLists.Find(id);
        if(funlist == null)
        {
            return HttpNotFound();
        }
        return View(funlist);
    }
    [HttpPost]
    public virtual ActionResult Edit(KFunList funlist)
    {
        if(ModelState.IsValid)
        {
            db.Entry(funlist).State = EntityState.Modified;
            db.SaveChanges();
            ViewBag.Message = "数据存储成功...";
        }
        else
        {
            ViewBag.Message = "数据输入有错误,数据存储失败...";
        }
        return View(funlist);
    }
    public virtual ActionResult Delete(string id = null)
    {
        KFunList funlist = db.KFunLists.Find(id);
        if(funlist == null)
        {
            return HttpNotFound();
        }
        return View(funlist);
    }
    [HttpPost,ActionName("Delete")]
```

```
            public virtual ActionResult DeleteConfirmed(string id)
            {
                KFunList funlist = db.KFunLists.Find(id);
                db.KFunLists.Remove(funlist);
                db.SaveChanges();
                return RedirectToAction("Index");
            }
            protected override void Dispose(bool disposing)
            {
                db.Dispose();
                base.Dispose(disposing);
            }
        }
```

控制器代码功能说明：

(1) 引入需要的类库。程序类库（Package）是 C#系统中类的集合，类是方法的集合，在方法使用之前需要引入相应的类库，以方便方法的调用；或者以"包.类库.方法"的方式调用。引入类库的语句是"using"。

(2) 定义数据库访问上下文实例变量。控制器的主要功能是在实体模型的基础上对数据进行处理，一方面需要从数据库中检索数据，加工处理，为视图提供数据源；另一方面，接受来自视图中的请求数据，加工处理，并将结果存储于数据库中。继承于数据库上下文类（DbContext）的用户自定义类"psjldb12Context"包含了系统实体模型对应的记录集合（即数据库中的表对象）名称定义，在数据处理时加以引入。定义数据库访问上文实例变量的语句为：

```
private psjldb12Context db = new psjldb12Context();
```

(3) 定义方法。根据数据处理的需要，增加相应的方法，以完成所要求的处理逻辑。对于记录操作，通用的处理功能有数据检索（Index）、新增记录（Create）、记录详细内容显示（Details）、记录删除（Delete）、记录数据编辑（Edit），其对应的默认方法（ActionResult）名称为"Index"、"Create"、"Details"、"Delete"、"Edit"。方法定义的基本格式为：

```
public virtual ActionResult Create()
{
    //数据处理语句
    return View();
}
```

（4）定义公用方法类库实例变量。"AppService.cs"是公用方法集合类库文件，位于命名空间"Models"，其中定义了系统常用的公用方法，例如，存取应用（Application）系统变量的用户自定义方法"SetApplication"。定义公用方法类库实例变量的语句为：

```
private AppService apps = new AppService();
```

上述有关控制器组成内容的说明，在后续章节可参照此处，将不再单独说明。

5.1.3 功能数据记录列表显示视图

功能项目属性数据记录显示视图的方法名为"Index"，对应的视图文件为"Index.cshtml"，其内容如下：

```
@model IEnumerable<psjlmvc4.Models.KFunList>
@{
    ViewBag.Title="系统功能管理";
}
<h2>@ViewBag.Title</h2>
<div class="noprint">
    @using(Ajax.BeginForm("Index","KFunList",new AjaxOptions{UpdateTargetId="dtable"}))
    {
        @Html.Label("功能编号:")
        @Html.TextBox("fcode")
        <input type="submit" value="查询"/>
        @Html.ActionLink("新建","Create",null,new{@class="alinkcss"})
        @Html.ActionLink("返回","Index","Home",new{Area=""},new{@class="alinkcss"})
        @Html.Label("功能数量:")@Model.Count();
    }
</div>
<table id="dtable">
    <thead>
    <tr>
        <th>
            @Html.DisplayNameFor(model=>model.funcode)
        </th>
        <th>
```

```
                @Html.DisplayNameFor(model => model.funname)
            </th>
            <th>
                @Html.DisplayNameFor(model => model.actionname)
            </th>
            <th>
                @Html.DisplayNameFor(model => model.controllername)
            </th>
            <th>
                @Html.DisplayNameFor(model => model.remark)
            </th>
            <th></th>
        </tr>
    </thead>
    <tbody>
@foreach(var item in Model) {
        <tr>
            <td>
                @Html.DisplayFor(modelItem => item.funcode)
            </td>
            <td>
                @Html.DisplayFor(modelItem => item.funname)
            </td>
            <td>
                @Html.DisplayFor(modelItem => item.actionname)
            </td>
            <td>
                @Html.DisplayFor(modelItem => item.controllername)
            </td>
            <td>
                @Html.DisplayFor(modelItem => item.remark)
            </td>
            <td>
                @Html.ActionLink("编辑","Edit",new{id=item.funcode}) |
                @Html.ActionLink("详细","Details",new{id=item.funcode}) |
                @Html.ActionLink("删除","Delete",new{id=item.funcode})
            </td>
        </tr>
}
    </tbody>
</table>
```

这是基于 Razor 语法引擎的 HTML 文件，在基于 C#语言为编程工具时，文件类型为"cshtml"，以后相同。

5.1.4 新增功能项目管理视图

实现新增功能项目管理的视图方法是"Create"，对应的视图文件名称是"Create.cshtml"，其代码内容如下：

```
@model psjlmvc4.Models.KFunList
@{
    ViewBag.Title = "新建设系统功能";
}
@using(Html.BeginForm()){
    @Html.ValidationSummary(true)
    <fieldset>
        <legend>@ViewBag.Title</legend>
        <div class="editor-label">
            @Html.LabelFor(model => model.funcode)
            @Html.EditorFor(model => model.funcode)
            @Html.ValidationMessageFor(model => model.funcode)
        </div>
        <div class="editor-label">
            @Html.LabelFor(model => model.funname)
            @Html.EditorFor(model => model.funname)
            @Html.ValidationMessageFor(model => model.funname)
        </div>
        <div class="editor-label">
            @Html.LabelFor(model => model.actionname)
            @Html.EditorFor(model => model.actionname)
            @Html.ValidationMessageFor(model => model.actionname)
        </div>
        <div class="editor-label">
            @Html.LabelFor(model => model.controllername)
            @Html.EditorFor(model => model.controllername)
            @Html.ValidationMessageFor(model => model.controllername)
        </div>
        <div class="editor-label">
            @Html.LabelFor(model => model.remark)
            @Html.EditorFor(model => model.remark)
            @Html.ValidationMessageFor(model => model.remark)
```

```
            </div>
            <hr/>
            <p>
                <input type="submit" value="确认存储"/>
                @Html.ActionLink("返回列表","Index","KFunList",null,new{@class="alinkcss"})
            </p>
        </fieldset>
}
@section Scripts{
    @Scripts.Render("~/bundles/jqueryval")
}
```

在控制器中对应同名的两个方法(多态),首先是无参数方法 Create(),为视图入口方法;第二是有模型参数(实体模型)传递的方法 Create(KFunList funlist),并以"HttpPost"方式调用。

5.1.5 功能项目详细内容显示视图

功能项目记录详细内容显示是以记录为单位,其方法名称为"Details",对应的视图文件名为"Details.cshtml",其代码内容如下:

```
@model psjlmvc4.Models.KFunList
@{
    ViewBag.Title="系统功能项目详细内容";
}
<fieldset>
    <legend>@ViewBag.Title</legend>
    <div class="display-label">
        @Html.DisplayNameFor(model=>model.funcode)
        @Html.DisplayFor(model=>model.funcode)
    </div>
    <div class="display-label">
        @Html.DisplayNameFor(model=>model.funname)
        @Html.DisplayFor(model=>model.funname)
    </div>
    <div class="display-label">
        @Html.DisplayNameFor(model=>model.actionname)
        @Html.DisplayFor(model=>model.actionname)
```

```
        </div>
        <div class = "display-label">
            @Html.DisplayNameFor(model = >model.controllername)
            @Html.DisplayFor(model = >model.controllername)
        </div>
        <div class = "display-label">
            @Html.DisplayNameFor(model = >model.remark)
            @Html.DisplayFor(model = >model.remark)
        </div>
    </fieldset>
    <p>
        @Html.ActionLink("返回功能列表","Index","KFunList",null,new{@class = "alinkc-ss"})
    </p>
```

5.1.6 功能项目记录数据编辑视图

功能项目记录数据编辑方法名称为 "Edit"，相应的视图文件名称为 "Edit.cshtml"，其代码内容如下：

```
@model psjlmvc4.Models.KFunList
@{
    ViewBag.Title = "系统功能信息编辑";
}
@using(Html.BeginForm()){
    @Html.ValidationSummary(true)
    <fieldset>
        <legend>@ViewBag.Title</legend>
        <div class = "editor-label">
            @Html.LabelFor(model = >model.funcode)
            @Html.DisplayFor(model = >model.funcode)
            @Html.HiddenFor(model = >model.funcode)
        </div>
        <hr/>
        <div class = "editor-label">
            @Html.LabelFor(model = >model.funname)
            @Html.EditorFor(model = >model.funname)
            @Html.ValidationMessageFor(model = >model.funname)
        </div>
```

```
            <div class="editor-label">
                @Html.LabelFor(model => model.actionname)
                @Html.EditorFor(model => model.actionname)
                @Html.ValidationMessageFor(model => model.actionname)
            </div>
            <div class="editor-label">
                @Html.LabelFor(model => model.controllername)
                @Html.EditorFor(model => model.controllername)
                @Html.ValidationMessageFor(model => model.controllername)
            </div>
            <div class="editor-label">
                @Html.LabelFor(model => model.remark)
                @Html.EditorFor(model => model.remark)
                @Html.ValidationMessageFor(model => model.remark)
            </div>
            <hr/>
            <p>
                <input type="submit" value="确认存储"/>
                @Html.ActionLink("返回列表","Index","KFunList",null,new{@class="alinkcss"})
                <span>@ViewBag.Message</span>
            </p>
        </fieldset>
}
@section Scripts{
    @Scripts.Render("~/bundles/jqueryval")
}
```

在控制器中对应同名的两个方法（多态），首先是单记录数据显示方法 Edit (string id = null)，为视图入口方法，其中，参数"string id = null"为字符变量定义（ID 记录主键）；第二是有模型参数（实体模型）传递的方法 Edit（KFunList funlist），并以"HttpPost"方式起行调用。

5.1.7 功能项目记录删除功能视图

系统功能项目记录删除功能实现的方法名称为"Delete"，对应的视图文件名称为"Delete.cshtml"，其代码内容如下：

```
@model psjlmvc4.Models.KFunList
@{
    ViewBag.Title = "删除系统功能";
}
<h2>确认删除吗?</h2>
<fieldset>
    <legend>@ViewBag.Title</legend>
    <div class="display-label">
        @Html.DisplayNameFor(model => model.funcode)
        @Html.DisplayFor(model => model.funcode)
    </div>
    <div class="display-label">
        @Html.DisplayNameFor(model => model.funname)
        @Html.DisplayFor(model => model.funname)
    </div>
    <div class="display-label">
        @Html.DisplayNameFor(model => model.remark)
        @Html.DisplayFor(model => model.remark)
    </div>
    <div class="display-label">
        @Html.DisplayNameFor(model => model.actionname)
        @Html.DisplayFor(model => model.actionname)
    </div>
    <div class="display-label">
        @Html.DisplayNameFor(model => model.controllername)
        @Html.DisplayFor(model => model.controllername)
    </div>
</fieldset>
@using(Html.BeginForm()){
    <p>
        <input type="submit" value="确认删除"/> |
        @Html.ActionLink("返回功能列表","Index",null,new{@class="alinkcss"})
    </p>
}
```

在控制器中对应同名的两个方法（多态），首先是单记录数据显示方法 Delete (string id = null)，为视图入口方法，其中，参数"string id = null"为字符变量定义（ID 记录主键）；第二个是有模型参数（实体模型）传递的方法 DeleteConfirmed (string id = null)，并以"HttpPost"方式调用（[HttpPost, ActionName ("Delete")]）。

5.2 系统角色管理

系统角色（Role）是系统用户类别分组的一种依据，不同类别的用户，可以分配不同的权限和规则，从而执行相应权限允许的功能（职责），完成对应的任务。在信息系统中，用户权限和功能的分配是通过所属角色的功能实现的。

系统角色实体模型名称是"KGroupList"，其定义内容存储于区域目录 KArea/Models 下的类定义集合文件"KModels.cs"中，对应的命名空间名称是"Models"。

5.2.1 系统角色管理控制器

系统角色管理控制器的类名为"KGroupListController"，继承于系统"Controller"类，相应的类文件名称为"KGroupListController.cs"，其代码内容如下：

```csharp
using psjlmvc4.Models;
using System;
using System.Data.Entity;
using System.Linq;
using System.Web.Mvc;
namespace psjlmvc4.Areas.KArea.Controllers
{
    public partial class KGroupListController:Controller
    {
        private psjldb12Context db = new psjldb12Context();
        private AppService apps = new AppService();
        public virtual ActionResult Index(FormCollection fc)
        {
            string gid = "",gname = "";
            int rps = 20;
            if(fc.AllKeys.Count() > 0)
            {
                apps.SetSession("grid_groupid",fc["grid_groupid"]);
                apps.SetSession("grid_groupname",fc["grid_groupname"]);
                apps.SetSession("grid_rowsperpage",fc["grid_rowsperpage"]);
            }
            gid = apps.GetSession("grid_groupid");
            gname = apps.GetSession("grid_groupname");
            rps = String.IsNullOrEmpty(apps.GetSession("grid_rowsperpage"))? rps: Convert.ToInt32(apps.GetSession("grid_rowsperpage"));
```

```csharp
            ViewBag.groupid = gid;
            ViewBag.groupname = gname;
            ViewBag.rowsperpage = rps;
            var rd = db.KGroupLists.Where(g => g.groupid.Contains(gid) && g.groupname.Contains(gname)).OrderBy(g => g.groupid);
            return View(rd.ToList());
        }
        public virtual ActionResult Details(string id = null)
        {
            KGroupList grouplist = db.KGroupLists.Find(id);
            if(grouplist == null)
            {
                return HttpNotFound();
            }
            return View(grouplist);
        }
        /// <summary>
        ///远程校验角色编号方法:CodeValidate
        /// </summary>
        /// <param name="groupid"></param>
        /// <returns>bool</returns>
        [HttpGet]
        public virtual ActionResult CodeValidate(string groupid = null)
        {
            var rd = db.KGroupLists.Find(groupid.ToUpper());
            bool exists = true;
            if(rd == null)
            {
                exists = false;
            }
            return Json(!exists, JsonRequestBehavior.AllowGet);
        }
        public virtual ActionResult Create()
        {
            ViewBag.Message = "根据提示输入数据...";
            return View();
        }
        [HttpPost]
        public virtual ActionResult Create(KGroupList grouplist)
        {
```

```
            if(ModelState.IsValid)
            {
                grouplist.groupid.ToUpper();
                db.KGroupLists.Add(grouplist);
                db.SaveChanges();
                ViewBag.Message ="数据存储完成...";
                //return RedirectToAction("Index");
            }
            else
            {
                ViewBag.Message ="数据有错误,存储失败...";
            }
            return View(grouplist);
        }
        public virtual ActionResult Edit(string id = null)
        {
            KGroupList grouplist = db.KGroupLists.Find(id);
            if(grouplist = = null)
            {
                return HttpNotFound();
            }
            if(id = ="A")
            {
                ViewBag.Nessage ="角色"+ grouplist.groupname +"是系统保留,不能编辑...";
                return RedirectToAction("Index");
            }
            return View(grouplist);
        }
        [HttpPost]
        public virtual ActionResult Edit(KGroupList grouplist)
        {
            if(ModelState.IsValid)
            {
                db.Entry(grouplist).State = EntityState.Modified;
                db.SaveChanges();
                ViewBag.Message ="存储成功...";
            }
            else
            {
```

```
                ViewBag.Message = "数据输入有错误!";
            }
            return View(grouplist);
        }
        public virtual ActionResult Delete(string id = null)
        {
            KGroupList grouplist = db.KGroupLists.Find(id);
            if(grouplist == null)
            {
                return HttpNotFound();
            }
            if(id == "A")
            {
           ViewBag.Nessage = "角色系统管理是系统保留,不能编辑...";
                return RedirectToAction("Index");
            }
            return View(grouplist);
        }
        [HttpPost,ActionName("Delete")]
        public virtual ActionResult DeleteConfirmed(string id)
        {
            KGroupList grouplist = db.KGroupLists.Find(id);
            db.KGroupLists.Remove(grouplist);
            db.SaveChanges();
            return RedirectToAction("Index");
        }
        protected override void Dispose(bool disposing)
        {
            db.Dispose();
            base.Dispose(disposing);
        }
    }
}
```

方法对应的显示视图文件存储于目录"Views/KGroupList"中。

5.2.2 角色数据记录列表显示视图

角色数据记录显示功能的方法名称为"Index",对应的视图文件为"Index.cshtml",其代码内容如下:

```
@model IEnumerable<psjlmvc4.Models.KGroupList>
@{
    ViewBag.Title = "系统角色管理";
    string idid = "";
    int i = 1;
    var grid = new WebGrid(source: Model,
        columnNames: new[] { "groupid", "groupname", "remark" },
        defaultSort: "groupid",
        rowsPerPage: (int)ViewBag.rowsperpage,
        canPage: true,
        canSort: true,
        ajaxUpdateContainerId: "glist",
        ajaxUpdateCallback: null,
        fieldNamePrefix: "grid_",
        pageFieldName: "pageIndex",
        selectionFieldName: "selectedRow",
        sortDirectionFieldName: "sortDirectionField",
        sortFieldName: "sortField");
}
<table id="glist" style="width:100%;">
    <caption><h2>@ViewBag.Title</h2></caption>
    <tr>
        <td style="text-align:left;
                   vertical-align:top;
                   width:70%;padding:0px;">
            @grid.GetHtml(tableStyle: "tableStyle",
                headerStyle: "headerStyle",
                selectedRowStyle: "selectedRowStyle",
                caption: "系统用户角色列表",
                alternatingRowStyle: "",
                columns: grid.Columns(
                    grid.Column(header: "序号", format: item => grid.PageIndex * grid.RowsPerPage + (i++), style: "selectColumnStyle"),
                    grid.Column("groupid", "角色编号", style: "fcode"),
                    grid.Column("groupname", "角色名称"),
                    grid.Column("remark", "角色说明", style: "fname", canSort: false),
                    grid.Column(format: item => item.GetSelectLink(), style: "selectColumnStyle")),
                mode: WebGridPagerModes.All,
                firstText: "第一页",
```

```
                previousText:"上一页",
                nextText:"下一页",
                lastText:"最后一页",
                numericLinksCount:5,
                htmlAttributes:new{id = "ghtml",onclick = ""})
        <div style = "padding:3px 20px 3px 20px;">
            记录数:@Html.Encode(grid.TotalRowCount)
            总页数:@Html.Encode(grid.PageCount)
            当前页:@Html.Encode(grid.PageIndex + 1)
            行索引:@Html.Encode(grid.SelectedIndex + 1)
        </div>
    </td>
    <td style = "text - align:left;
            vertical - align:top"class = "noprint">
        操作选项<hr/>
        @using(Html.BeginForm("Index",null,FormMethod.Post))
        {
            @Html.Label("角色编号")
            @Html.TextBox("grid_groupid",(string)ViewBag.groupid,new{style = "width:200px;"})<br/>
            @Html.Label("角色名称")
            @Html.TextBox("grid_groupname",(string)ViewBag.groupname,new{style = "width:200px;"})<br/>
            @Html.Label("每页行数")
            @Html.TextBox("grid_rowsperpage",(int)ViewBag.rowsperpage,new{style = "width:200px;"})<br/>
            <hr/>
            <input type = "submit"value = "查询"/>
            @Html.ActionLink("新增","Create",null,new{@class = "alinkcss"})
            <button onclick = "window.print();return false;">打印</button>
            @Html.ActionLink("返回","Index","Home",new{Area = ""},new{@class = "alinkcss"})
        }
        <hr/>
        @if(grid.HasSelection)
        {
            idid = Convert.ToString(grid.SelectedRow.ElementAt(0));
            @Html.Encode(grid.SelectedRow.ElementAt(0));
            @Html.Raw(" - ");
            @Html.Encode(grid.SelectedRow.ElementAt(1));
```

```
                    @Html.Raw("-");
                    @Html.Encode(grid.SelectedRow.ElementAt(2));
                    @Html.Raw("-");
                    @Html.Encode(grid.SelectedIndex+1);
                    <hr/>
                    if(idid!="A")
                    {
                        @Html.ActionLink("编辑","Edit",new{id=idid},new{@class="alinkcss"})
                        @Html.ActionLink("详细","Details",new{id=idid},new{@class="alinkcss"})
                        @Html.ActionLink("删除","Delete",new{id=idid},new{@class="alinkcss"})
                    }
                    else
                    {
                        @:<span>"系统管理员"角色不能删除!</span>
                    }
                </td>
            </tr>
        </table>
```

"WebGrid"是MVC内置帮助器,通过其"GetHtml"扩展方法以表格方式显示记录,可以实现分页、排序、字段选择、显示格式化等功能。

5.2.3 新增角色功能视图

系统角色新增功能实现的方法名称是"Create",与其对应的视图文件名称为"Create.cshtml",其代码内容如下:

```
@model psjlmvc4.Models.KGroupList
@{
    ViewBag.Title="新增角色";
}
@using(Html.BeginForm()){
    @Html.ValidationSummary(true)
    <fieldset>
        <legend>@ViewBag.Title</legend>
        <div class="editor-label">
```

```
            @Html.LabelFor(model => model.groupid)
            @Html.EditorFor(model => model.groupid)
            @Html.ValidationMessageFor(model => model.groupid)
        </div>
        <div class="editor-label">
            @Html.LabelFor(model => model.groupname)
            @Html.EditorFor(model => model.groupname)
            @Html.ValidationMessageFor(model => model.groupname)
        </div>
        <div class="editor-label">
            @Html.LabelFor(model => model.remark)
            @Html.EditorFor(model => model.remark)
            @Html.ValidationMessageFor(model => model.remark)
        </div>
        <hr/>
        <p>
            <input type="submit" value="确认存储"/>
            @Html.ActionLink("返回列表","Index",null,new{@class="alinkcss"})
        </p>
        <hr/>
        <div>操作提示:</div>
        <ol>
<li>系统角色编号是一位大写英文字母,例如,"A"--"系统管理"角色;</li>
<li>系统角色名称力求简洁明确,含义准确,长度不能超过20个字;</li>
<li>角色编号一经定义存储,不能修改;</li>
<li>通过删除某个系统角色,再新增,写成编号修改。</li>
        </ol>
        <hr/>
        <span>@ViewBag.Message</span>
    </fieldset>
}
@section Scripts{
    @Scripts.Render("~/bundles/jqueryval")
}
```

5.2.4 角色数据记录详细内容显示视图

角色数据记录详细内容显示是以记录为单位,其方法名称为"Details",对应的视图文件名为"Details.cshtml",其代码内容如下:

```
@model psjlmvc4.Models.KGroupList
@{
    ViewBag.Title = "系统角色记录详细内容";
}
<table id="tt-table">
    <caption>@ViewBag.Title</caption>
    <tr>
        <td class="td-label">
            @Html.DisplayNameFor(model => model.groupid)
        </td>
        <td class="td-field">
            @Html.DisplayFor(model => model.groupid)
        </td>
    </tr>
    <tr>
        <td class="td-label">
            @Html.DisplayNameFor(model => model.groupname)
        </td>
        <td class="td-field">
            @Html.DisplayFor(model => model.groupname)
        </td>
    </tr>
    <tr>
        <td class="td-label">
            @Html.DisplayNameFor(model => model.remark)
        </td>
        <td class="td-field">
            @Html.DisplayFor(model => model.remark)
        </td>
    </tr>
</table>
<p>
    @Html.ActionLink("返回列表","Index",null,new{@class="alinkcss"})
</p>
```

5.2.5 角色数据记录编辑功能视图

角色数据记录编辑功能的方法名称为"Edit",相应的视图文件名称为"Edit.cshtml",其代码内容如下:

```
@model psjlmvc4.Models.KGroupList
@{
    ViewBag.Title = "系统角色编辑";
}
@using(Html.BeginForm())
{
    @Html.ValidationSummary(true)
    <fieldset>
        <legend>@ViewBag.Title</legend>
        <div>
            @Html.LabelFor(model => model.groupid)
            @Html.DisplayFor(model => model.groupid)
            @Html.HiddenFor(model => model.groupid)
        </div>
        <hr/>
        <div class="editor-label">
            @Html.LabelFor(model => model.groupname)
            @Html.EditorFor(model => model.groupname)
            @Html.ValidationMessageFor(model => model.groupname)
        </div>
        <div class="editor-label">
            @Html.LabelFor(model => model.remark)
            @Html.EditorFor(model => model.remark)
            @Html.ValidationMessageFor(model => model.remark)
        </div>
        <hr/>
        <p>
            <input type="submit" value="确认存储"/>
            @Html.ActionLink("返回列表","Index",null,new{@class="alinkcss"})
        </p>
        <span>@ViewBag.ErrorMessage</span>
    </fieldset>
}
@section Scripts{
    @Scripts.Render("~/bundles/jqueryval")
}
```

5.2.6 角色记录删除功能视图

角色记录删除功能实现的方法名称为"Delete",对应的视图名称为

"Delete.cshtml"，其代码内容如下：

```
@model psjlmvc4.Models.KGroupList
@{
    ViewBag.Title = "系统用户角色删除";
}
<h2>确认删除此记录吗？</h2>
<fieldset>
    <legend>@ViewBag.Title</legend>
    <div class="display-label">
        @Html.DisplayNameFor(model => model.groupid)
        @Html.DisplayFor(model => model.groupid)
    </div>
    <div class="display-label">
        @Html.DisplayNameFor(model => model.groupname)
        @Html.DisplayFor(model => model.groupname)
    </div>
    <div class="display-label">
        @Html.DisplayNameFor(model => model.remark)
        @Html.DisplayFor(model => model.remark)
    </div>
</fieldset>
@using(Html.BeginForm()){
    <p>
        <input type="submit" value="确认删除"/>
        @Html.ActionLink("返回列表","Index",null,new{@class="alinkcss"})
    </p>}
```

5.3 用户角色分配

注册的系统用户需要分配相应的系统角色，以便系统确定用户的权限和功能。不同的用户可以分配同一个角色，对于角色实体中的一个实例，可以对应不同的多个用户实例，因此系统用户（实体为 KUserStaff）和系统角色（实体为 KGroupList）是多对一（反过来是一对多）的关系。

5.3.1 一对多关系定义

在系统用户实体属性定义中有属性"groupid"，其定义内容如下：

```
[Display(Name ="角色",GroupName ="User",Order =1)]
public string groupid{get;set;}//from KGroupList
```

"groupid"属性和角色实体中的主键"groupid"性质相同,是借用,属于外键,以保持和实体"KGroupList"的联系。一对多实体模型关系在实体模型定义中的方法如下:

(1) 在系统角色(主实体)模型(KGroupList)中定义基于系统用户实体的虚拟集合变量 LUserStaffs,定义语句如下:

```
public virtual ICollection < LUserStaff > LUserStaffs{get;set;}
```

(2) 在系统用户依赖实体模型(LUserStaff)中定义基于系统角色实体的虚拟类变量 KGroupList,定义语句如下:

```
public virtual KGroupList KGroupList{get;set;}
```

这样,系统角色实体和系统用户实体之间就建立了一对多的关联关系。

5.3.2 系统角色记录检索

一对多关联关系建立后,在用户管理实现的控制器方法中可检索角色记录信息集合,这个集合中包含数据项"groupid"和"groupname"。"groupname"用于列表显示的项目内容,"groupid"用于存储于实体属性"groupid"的值。

控制器方法中,记录检索的语句代码如下:

```
ViewBag.groupid = new SelectList(db.KGroupLists,"groupid","groupname",ka.groupid);
```

其中,"db.KGroupLists"是角色记录集合,"ka.groupid"是当前默认值,选取的数据项目是"groupid"和"groupname",语句同时出现在方法"Create"和"Edit"中。

5.3.3 视图中实现用户角色选择

系统用户角色分配是通过给用户实体属性"groupid"定义相应的值来实现的,这个值是根据系统角色实体记录值选取,完成选取的操作在视图中进行。

在"Create"和"EditUser"两个视图中,完成此功能的 HTML 表单控件代码内容如下:

```
……
    < tr >
        < td class = "td – label" >
            @ Html. LabelFor( model = > model. groupid)
        </td >
        < td class = "td – field" >
            @ Html. DropDownList("groupid", String. Empty)
        </td >
    </tr >
……
```

其中,"groupid"是系统角色记录数据集合,由"ViewBag. groupid"传递过来;"String. Empty"是默认选项值。运行时通过下拉列表选取,如图 5 – 1 所示。

其中,"所在组别"即为用户的角色定义项目,下拉列表框是角色记录集合。

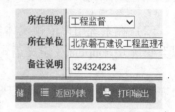

图 5 – 1 用户角色分配选择视图

5.4 角色功能分配

系统功能(二级功能)管理是根据当前登录用户的角色所拥有的权限进行分类管理,即不同的用户,角色不同,权限不同,可操作的功能也不同;不同的用户,角色相同(同组),则权限相同,可操作的功能也相同。用户角色定义,决定了用户的权限和可操作功能,关键是为系统中每一类角色根据权限合理分配相应的功能。

5.4.1 系统角色实体与系统功能实体的关系

系统角色实体数据模型类名称是"KGroupList",系统功能实体数据模型类名称是"KFunList"。一个角色可以拥有多个功能,反之,一个功能可以被多个角色拥有,因此,二者为"多对多"的逻辑关系,如图 5 – 2 所示。

系统角色实体数据模型的主键(PK)是"groupid",系统功能实体数据模型的主键(PK)是"funcode",以二者为外键,由此建立"角色功能"的关系实体"KGroupFun",并增加属性"id"作为其主键(其定义为 public int id {get; set;}),这样,角色和功能实体之间的"多对多"关系就转换为角色实体和角色功能实体、功能实体和角色功能实体两个"一对多"的关系,如图 5 – 3 所示。

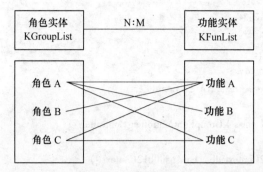

图5-2 角色和功能实体"多对多"的逻辑关系

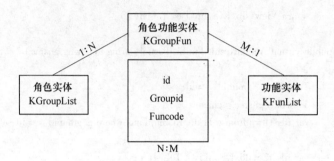

图5-3 角色和功能实体"一对多"的逻辑关系

相应的模型定义中的关系定义如下：
（1）角色模型：

```
public virtual ICollection < KGroupFun > KGroupFuns{get;set;}
```

（2）功能模型：

```
public virtual ICollection < KGroupFun > KGroupFuns{get;set;}
```

（3）角色功能模型：

```
public virtual KGroupList KGroupList{get;set;}
public virtual KFunList KFunList{get;set;}
```

5.4.2 角色—功能分配功能实现的控制器

角色—功能分配功能实现的控制器名称为"KGroupFunController"，对应的控制类文件名称为"KGroupFunController.cs"，其代码内容如下：

```csharp
using psjlmvc4.Models;
using System;
using System.Linq;
using System.Web.Mvc;
namespace psjlmvc4.Areas.KArea.Controllers
{
    public partial class KGroupFunController:Controller
    {
        psjldb12Context db = new psjldb12Context();
        public virtual ActionResult Index()
        {
            return View(db.KGroupLists.ToList());
        }
        public virtual ActionResult FunListPartial(string id,string gname)
        {
            TempData["groupid"] = id;
            ViewBag.gname = gname;
            var rdset1 = (from g in db.KGroupFuns where g.groupid == id select g).ToList();
            var rdset2 = db.KFunLists.ToList();
            int cn = 0;
            foreach(var item in rdset2)
            {
                var rd = from r in rdset1 where r.funcode == item.funcode select r;
                if(rd.Count() > 0)
                {
                    item.yesno = true;
                }
                else
                {
                    item.yesno = false;
                }
                cn ++;
                UpdateModel(item);
            }
            ViewBag.cn = cn;
            return PartialView(rdset2);
        }
        [HttpPost]
        public virtual ActionResult SetGroupFun(string id,string[] yesno)
        {
```

```
            try
            {
                ViewBag.groupid = id;
        var rd = from g in db.KGroupFuns where g.groupid = = id select g;
                int dels = 0;
                foreach(var item in rd.ToList())
                {
                    db.KGroupFuns.Remove(item);
                    dels + + ;
                }
                db.SaveChanges();
                int adds = 0;
                foreach(var item in yesno)
                {
                    KGroupFun gf = new KGroupFun();
                    gf.groupid = id;
                    gf.funcode = item;
                    db.KGroupFuns.Add(gf);
                    adds + + ;
                }
                db.SaveChanges();
    ViewBag.message = "删除功能数:" + dels + ";增加功能数:" + adds;
            }
            catch(Exception ee)
            {
                ViewBag.message = ee.Message + "角色功能设置失败!";
            }
            return JavaScript("alert('" + ViewBag.message + "');");
        }
    }
```

控制器由三个方法组成，分别为："Index"，显示角色集合记录，由同名的视图实现；"FunListPartial"，显示选中角色对应的功能集合记录，由同名的分部视图实现；"SetGroupFun"，完成功能设置并将结果存储于数据库，通过JavaScript 显示执行结果。

5.4.3 系统角色记录显示视图

系统角色记录显示的方法名称是"Index"，对应的视图文件名称是"Index.cshtml"，其代码内容如下：

```
@model IEnumerable<psjlmvc4.Models.KGroupList>
@{
    ViewBag.Title = "系统角色功能设置";
}
<h2>@ViewBag.Title</h2>
<div class="row">
    <div class="col-sm-5">
        <table id="dtable">
            <caption><h2>系统角色</h2></caption>
            <tr>
                <th>@Html.DisplayNameFor(model => model.groupid)</th>
                <th>@Html.DisplayNameFor(model => model.groupname)</th>
                <th>@Html.DisplayNameFor(model => model.remark)</th>
                <th></th>
            </tr>
            @foreach(var item in Model)
            {
                <tr>
                    <td>@Html.DisplayFor(modelitem => item.groupid)</td>
                    <td>@Html.DisplayFor(modelitem => item.groupname)</td>
                    <td>@Html.DisplayFor(modelitem => item.remark)</td>
                    <td>
                        @Ajax.ActionLink("功能设置","FunListPartial","KGroupFun", new{id=item.groupid, gname=item.groupname},
                            new AjaxOptions{UpdateTargetId="div-funlist"})
                    </td>
                </tr>
            }
        </table>
        <p>
            @Html.ActionLink("返回","Index","Home", new{Area=""}, new{@class="btn btn-primary glyphicon glyphicon-home"})
        </p>
    </div>
    <div class="col-sm-7">
        <div id="div-funlist">
            功能记录显示
        </div>
    </div>
</div>
```

"功能记录显示"是通过 Ajax 调用、局部更新后功能记录内容显示，Ajax 调用是由"@Ajax.ActionLink("功能设置","FunListPartial","KGroupFun", new {id = item.groupid, gname = item.groupname}, new AjaxOptions {UpdateTargetId ="div-funlist"})"超级链接实现。

5.4.4 功能记录显示的局部视图

功能记录显示的局部视图在控制器中的方法名称是"FunListPartial"，对应的视图文件名称是"FunListPartial.cshtml"，其代码内容如下：

```
@model IEnumerable<psjlmvc4.Models.KFunList>
@{
    ViewBag.Title ="角色功能设置";
    var grid = new WebGrid(Model,defaultSort:"funcode",canPage:false,canSort:false);
}
@using(Ajax.BeginForm("SetGroupFun","KGroupFun",new{id=@TempData["groupid"]},new AjaxOptions{}))
{
    <h2>"@ViewBag.gname"@ViewBag.Title</h2>
    @grid.GetHtml(tableStyle:"tableStyle",headerStyle:"headerStyle",selectedRowStyle:"selectedRowStyle",
        columns:grid.Columns(
            grid.Column(format:@<input name="yesno"type="checkbox" value="@item.funcode"checked="@item.yesno"style="width:20px;"/>),
            grid.Column("funcode","编号",style:"fcode"),
            grid.Column("funname","功能名称"),
            grid.Column("remark","功能说明")))
    <p>
        <button id="all"type="button"class="btn btn-primary glyphicon glyphicon-ok-circle">全选</button>
        <button id="clear"type="button"class="btn btn-primary glyphicon glyphicon-remove-circle">全清</button>
        <button id="ok"type="submit"class="btn btn-primary glyphicon glyphicon-save">确认</button>
        @Html.Encode("功能数量:"+@ViewBag.cn)
    </p>
}
<script type="text/javascript">
    $(document).ready(function(){
        $("#all").click(function(){
            $("input[type='checkbox']").prop("checked",true);
```

```
        });
        $("#clear").click(function(){
            $("input[type='checkbox']").prop("checked",false);
        });
    })
</script>
```

在视图中，利用 jQuery 方法实现了所列功能的"全清"、"全选"选择功能，"确认"后将选择结果以 Ajax 方式传递到方法"SetGroupFun"进行处理（删除后增加）并存储，"SetGroupFun"以 JavaScript 方式返回处理结果，没有对应的视图文件。

5.5 用户登录与动态功能导航实现

系统定义的用户通过登录过程，确定其在系统中的角色，再由角色检索相应的功能，从而实现用户功能导航。

5.5.1 系统用户登录方法

系统用户登录实现的方法名称是"Loging"，所在控制器名称是"HomeController"，有登录视图显示和登录信息提交处理两个方法模态，其代码内容如下：

```
#region 系统用户登录管理 ********************
/// <summary>
///用户登录第一次启用此方法
/// </summary>
/// <returns> </returns>
[AllowAnonymous]
public virtual ActionResult Loging(string returnUrl)
{
    ViewBag.Message ="请输入正确的用户标识和用户密码!" + returnUrl;
    return View();
}
/// <summary>
///用户输入登录信息并提交后所启用的方法
/// </summary>
/// <param name ="fc"> </param>
/// <returns> </returns>
```

```csharp
[HttpPost]
[AllowAnonymous]
public virtual ActionResult Loging(KUserLoging ulm, string returnUrl)
{
    string uid = Convert.ToString(ulm.userid);
    string pwd = Convert.ToString(ulm.password);
    var rd = db.LUserStaffs.SingleOrDefault(u => u.userid == uid &&
                                                  u.password == pwd);
    if(rd == null)
    {
        ViewBag.Message = "登录失败！用户名称或口令输入有误！";
        return View(ulm);
    }
    //代码进行到此,说明登录成功
    rd.logincount += 1;
    db.SaveChanges();//更新登录次数
    #region 保存变量,供其他应用(页面)使用
    apps.SetSession("userid", uid);
    Session.Add("fullname", rd.fullname);
    apps.SetAppState("usercount", Convert.ToString(OnlineCount(uid, true)));
    //用户角色信息
    if(String.IsNullOrEmpty(rd.groupid))
    {
        Session.Add("groupid", "");
        Session.Add("groupname", "");
    }
    else
    {
        Session.Add("groupid", rd.groupid ?? "");
        Session.Add("groupname", rd.KGroupList.groupname ?? "");
    }
    //用户单位和单位类型信息
    if(String.IsNullOrEmpty(rd.unitcode))
    {
        Session.Add("unitcode", "");
        Session.Add("unitname", "");
        Session.Add("unittypename", "");
```

```
        }
        else
        {
                Session.Add("unitcode",rd.unitcode??"");
                Session.Add("unitname",rd.LUnitList.unitname??"");
        Session.Add("unittypename",rd.LUnitList.LUnitType.unittypename??"");
        }
        #endregion 保存变量结束
        //修改系统登录信息
        FormsAuthentication.SetAuthCookie(uid,true,
                        FormsAuthentication.FormsCookiePath);
        #region 写入日志
        KLoginList ul = new KLoginList();
        ul.userid = uid;
        ul.event1 = apps.GetSession("groupname");
        ul.logintime = DateTime.Now;
        ul.actiontype = "in";
        ul.hostname = Dns.GetHostName();
        ul.hostip = Dns.GetHostAddresses(Dns.GetHostName())[2].ToString();
        db.KLoginLists.Add(ul);
        db.SaveChanges();
        #endregion 写入日志结束
        if(!String.IsNullOrEmpty(returnUrl))
        {
                return Redirect(returnUrl);
        }
        else
        {
                return RedirectToAction("Index","Home");
        }
    }
}
#endregion 系统用户登录管理结束
```

登录成功后，方法所完成的任务分别是修改登录次数、确定用户角色、确定用户相关属性、记录用户登录信息，然后返回系统主页。

5.5.2 系统用户登录视图

系统用户登录实现的视图名称是"Loging.cshtml"，其代码内容如下：

```
@model psjlmvc4.Models.KUserLoging
@{
    ViewBag.Title = "系统用户切换";
}
<div class = "center-block">
    @using(Html.BeginForm())
    {
        <div class = "panel panel-primary" style = "width:600px;
            text-align:center;margin:auto;">
            <div class = "panel-heading" style = "text-align:left;">
                @ViewBag.Title
            </div>
            <div class = "panel-body">
                <div>
                    @Html.LabelFor(m => m.userid)
                    @Html.TextBoxFor(m => m.userid, new { placeholder = "请输入用户标识...", style = "opacity:0.5;"})
                    <span class = "message"> * </span>
                </div>
                <br/>
                <div>
                    @Html.LabelFor(m => m.password)
                    @Html.PasswordFor(m => m.password, new { placeholder = "请输入用户密码...", style = "opacity:0.5;"})
                    <span class = "message"> * </span>
                </div>
                <hr/>
                <div>
                    @Html.LabelFor(m => m.rememberme)
                    @Html.CheckBoxFor(m => m.rememberme)
                </div>
                <hr/>
                <div>
                    @Html.LabelFor(m => m.randinput)
                    @Html.TextBoxFor(m => m.randinput)
                    @Html.ValidationMessageFor(m => m.randinput)
                </div>
                <div>
                    @Html.LabelFor(m => m.randcode)
                    @Html.DisplayFor(m => m.randcode)
```

```
                    @Html.HiddenFor(m => m.randcode)
                </div>
            </div>
            <div class="panel-footer">
                <button type="submit" class="btn btn-primary glyphicon glyphicon-adjust">用户切换</button>
                @Html.ActionLink("返回主页","Index",null,new{@class="btn btn-primary glyphicon glyphicon-home"})
                <hr/>
                <div class="message">@ViewBag.Message</div>
            </div>
        </div>
    }
</div>
@section Scripts{
    @Scripts.Render("~/bundles/jqueryval")
}
```

其中包括验证码输入和比较的代码。

5.5.3 系统主（一级）功能导航

系统顶层（一级）功能导航内容在系统主页中直接全部显示，由对应的链接导航方式通过功能检索和局部视图方式显示相应的子功能。一级功能导航内容显示实现的视图是"Index.cshtml"，其代码内容如下：

```
@model IEnumerable<psjlmvc4.Models.KFunList>
@{
    ViewBag.Title = "系统主页";
    string fc = Session["funcode"]==null?"A":
                Session["funcode"].ToString();
}
<div class="row" style="font-size:large;">
    <div class="col-sm-8" style="border-right:4px double red;">
        <ul class="list-group list-inline" id="ul-mainmenu">
            @foreach(var item in Model)
            {
                <li class="list-group-item">
```

```
                        @Ajax.ActionLink(item.funname,"SubMenuListPartial","Home",
new{fcode=item.funcode},
                            new AjaxOptions{UpdateTargetId="div-submenulist"})
                    </li>
                }
            </ul>
        </div>
        <div class="col-sm-4">
            <h2>项目子功能</h2>
            <div id="div-submenulist" style="overflow-y:auto;height:300px;text-align:left;">
                @{Html.RenderAction("SubMenuListPartial","Home",new{fcode=fc});}
            </div>
        </div>
    </div>
</div>
<!--***********************************************-->
<style type="text/css">
    #ul-mainmenu li a{
        width:200px;
        display:inline-block;
    }
</style>
<script type="text/javascript">
    $(document).ready(function(){
        var licolor=$("#ul-mainmenu li a").css("background-color");
        $("#ul-mainmenu li a").click(function(){
            $("#ul-mainmenu li a").css("background-color",licolor);
            $(this).css("background-color","red");
        })
    })
</script>
```

因为在系统用户登录时，系统用户的角色编号已经通过变量Session["groupid"]进行了保存；在此，可以通过方法参数将当前选择的一级功能编号传递到子功能检索相应的处理方法，就可以实现相应子功能导航内容的显示。

控制器"HomeController"中调用主页的方法"Index"的代码内容如下：

```
/// < summary >
///系统门户首页
public virtual ActionResult Index( )
{
    ViewBag. Message = "系统门户首页";
    var rd = db. KFunLists. Where( k = > k. funcode. Trim( ). Length = = 1);
    return View( rd. ToList( ) );
}
```

一级功能导航内容显示格式如图 5 – 4 所示。

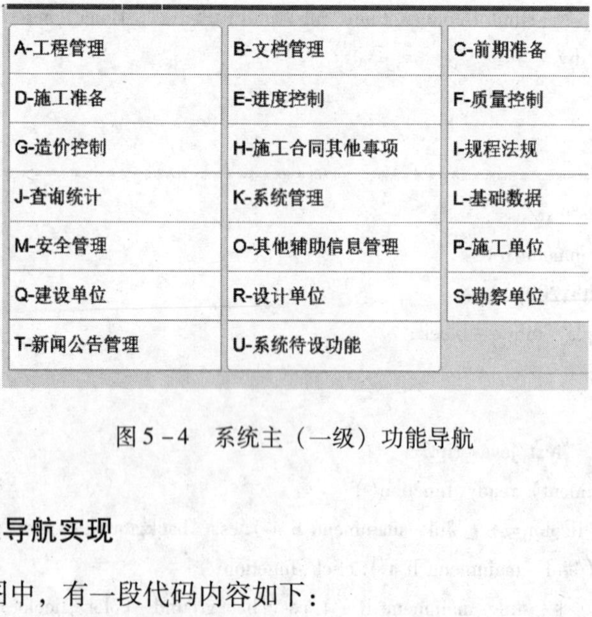

A-工程管理	B-文档管理	C-前期准备
D-施工准备	E-进度控制	F-质量控制
G-造价控制	H-施工合同其他事项	I-规程法规
J-查询统计	K-系统管理	L-基础数据
M-安全管理	O-其他辅助信息管理	P-施工单位
Q-建设单位	R-设计单位	S-勘察单位
T-新闻公告管理	U-系统待设功能	

图 5 – 4　系统主（一级）功能导航

5.5.4　子功能导航实现

在主页视图中，有一段代码内容如下：

```
@ Ajax. ActionLink( item. funname,"SubMenuListPartial","Home",
    new{ fcode = item. funcode} ,
    new AjaxOptions{ UpdateTargetId = "div – submenulist"} )
```

这是主（一级）功能导航链接激活时执行的代码，其作用是通过传递当前主功能代码（fcode = item. funcode）到方法"SubMenuListPartial"，检索相应子功能导航记录，再以局部更新的方式通过局部视图显示子功能导航记录内容（Ajax 调用），以实现系统二级功能导航系统。"SubMenuListPartial"方法在控制器"HomeController"中，其相应的代码内容如下：

```csharp
//显示子菜单功能(分部视图)
public ActionResult SubMenuListPartial(string fcode = "")
{
    apps.SetSession("funcode", fcode);
    var gid = apps.GetSession("groupid");
    var rd = from g in db.KGroupFuns
             where g.groupid == gid && g.funcode.Substring(0,1) == fcode &&
                 g.funcode.Length > 1
             from f in db.KFunLists
             where f.funcode == g.funcode
             orderby f.funcode
             select f;
    return PartialView(rd.ToList());
}
```

注意，方法返回的视图类型是局部视图（PartialView）。

5.5.5 子功能导航内容显示的局部视图

子功能导航内容显示的局部（分部）视图名称是"SubMenuListPartial.cshtml"，其代码内容如下：

```html
@model IEnumerable<psjlmvc4.Models.KFunList>
<div style="text-align:center;font-family:'Microsoft YaHei';
           font-size:20px;background-color:#7db9e8;">
    @ViewBag.mtitle
</div>
<ul id="submenu">
    @foreach(var item in Model)
    {
        <li>
            @Html.ActionLink(item.funname, item.actionname,
                item.controllername,
                new{ Area = @item.funcode.Substring(0,1) + "Area"}, null)
        </li>
    }
</ul>
```

工程管理主功能对应的子功能导航内容显示如图 5-5 所示。

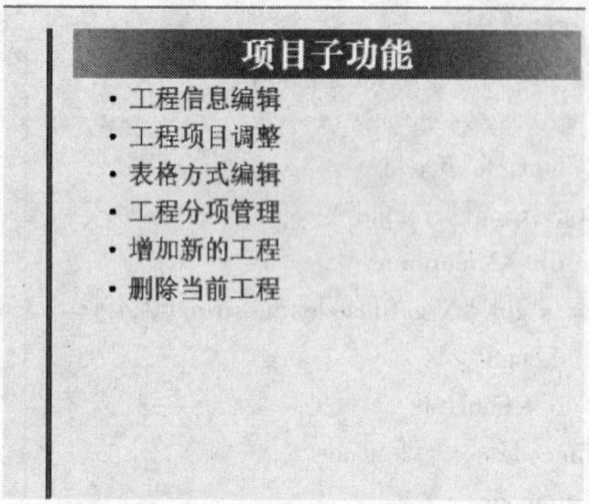

项目子功能
- 工程信息编辑
- 工程项目调整
- 表格方式编辑
- 工程分项管理
- 增加新的工程
- 删除当前工程

图 5-5　工程管理主功能对应的子功能导航运行示意图

本章小结

本章全面系统地介绍了系统功能导航实现过程所涉及的内容，包括用户管理、功能管理、角色管理、角色功能分配管理，以及导航实现有关的用户登录、主功能导航、子功能导航所需要的方法、模型和视图。

6 CRUD 模板设计

ASP.NET MVC 框架的应用系统开发的核心技术是基于实体框架（EF）结构的数据模型设计，这对构建 ASP.NET 应用程序的开发人员来说非常有吸引力。对数据库的常规操作，是通过对数据模型进行相应的操作来实现的，这些常规操作有数据检索（Retrieve）、记录增加（Create）、记录修改（Update）、记录删除（Delete），即所谓的 CRUD 操作。在原理和组成结构上，所有实体模型的 CRUD 在处理逻辑上是相似的，因此，CRUD 操作生成的模板是减少程序开发工作量、提高开发速度的关键。利用 CRUD 模板，可以快速自动生成实体模型的 CRUD 操作管理功能所需要的控制器和视图，并且开发人员还可以根据需要个性化修改模板，以适应自己的应用要求。

现以系统角色（KGroupList）实体模型管理所需要的控制器和相应视图的生成过程应用实例来说明 ASP.NET MVC 系统中 CRUD 模板（包括控制器和视图）的应用原理，根据"建设工程监理信息系统"项目开发的需要自定义模板内容。

本章主要内容包括：
6.1 CRUD 控制器模板应用实例
6.2 CRUD 视图模板应用实例
6.3 ASP.NET MVC 系统自有 CRUD 模板
6.4 自定义 CRUD 模板

6.1 CRUD 控制器模板应用实例

在 ASP.NET MVC 框架下，控制器是实体模型和模型数据表现视图之间的联系控制操作类，包含从数据库检索数据、处理数据、传递数据、接受数据、存储数据于数据库等功能实现的方法，这些方法就是所谓的 CRUD 操作。对于实体模型而言，CRUD 管理方法的模式结构类似，处理过程统一，这就是基于实体模型驱动方式的数据处理开发的优势，控制器就是 CRUD 处理方法的统一体现，控制器模板是提高开发速度的关键。

6.1.1 实体模型与数据库表的对应关系

系统角色实体模型具体定义如下:

```csharp
#region 4. KGroupList:系统角色
public class KGroupList
{
    public KGroupList()
    {
        this.LUserStaffs = new HashSet<LUserStaff>();
    }
    [Key]
    [Display(Name="角色序号")]
    [Required()]
    [Remote("CodeValidate","KGroupList",
                ErrorMessage="此用户角色已经存在")]
    [RegularExpression("^[A-Z]{1}$",
                ErrorMessage="{0}必须是一位大写字母")]
    public string groupid{get;set;}
    [Display(Name="角色名称")]
    [StringLength(50,ErrorMessage="{0}的长度不能超过{1}!")]
    public string groupname{get;set;}
    [Display(Name="备注说明")]
    [StringLength(500,ErrorMessage="{0}的长度不能超过{1}!")]
    public string remark{get;set;}
    public virtual ICollection<LUserStaff> LUserStaffs{get;set;}
    public virtual ICollection<KGroupFun> KGroupFuns{get;set;}
}
#endregion 4. KGroupList:系统角色结束
```

在模型中,共定义了三个属性,分别为"groupid"、"groupname"和"remark",其中"groupid"为模型的主键(由[Key]验证属性说明),模型相对比较简单。

模型并不存储数据,其数据是存储在数据库中与此模型对应的表(Table)中,根据约定在先的原则,模型"KGroupList"在数据库中对应的表的名称是"KGroupLists"(模型名称的复数形式),如图6-1所示。

角色实体模型"KGroupList"与数据库中对应的表"KGroupLists"之间的数据通讯和交换是通过定义在数据库连接上下文类文件"psjldb12Context.cs"中的集合对象定义语句"public DbSet<KGroupList> KGroupLists {get; set;}"所建立的关系而实现

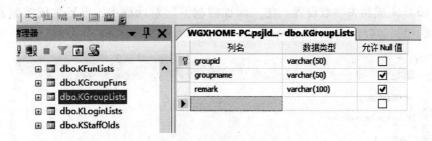

图6-1 模型与数据库中对应的表

的,这种对应要求模型中的实体属性定义和表对象中的属性定义保持一致。

6.1.2 建立 CRUD 控制器

有了模型、表和模型与表的上下文关联关系之后,利用 VS 系统提供的 CRUD 控制器模板就可以很容易地生成基于特定模型的 CRUD 操作的控制器。其建立过程及步骤如下:

(1)定位控制器存储目录。控制器默认存储目录是"Controllers"。定位到控制器存储目录"Controllers",此处控制器存储的目录是"Areas/KArea/Controllers"。

(2)添加控制器。在目录名"Controllers"上单击右键,在菜单中选择"添加(D)/控制器",打开"添加控制器"选择对话框,如图6-2所示。

图6-2 添加控制器对话框

（3）完成相关参数设置。在"添加控制器"对话框中，设置生成控制器所需要的相关参数，如图6-3所示。

图6-3　添加控制器参数设置对话框

参数设置完成后，单击"添加"，根据 CRUD 现有模板，系统开始生成相应的 CRUD 操作功能管理实现的控制器和对应的视图。

6.1.3　CRUD 控制器代码内容组成

根据上述参数，系统根据现有 CRUD 模板所生成的实际 CRUD 控制器的代码内容如下：

```
using psjlmvc4.Models;
using System;
using System.Data.Entity;
using System.Linq;
using System.Web.Mvc;
namespace psjlmvc4.Areas.KArea.Controllers
{
    public partial class KGroupListController:Controller
    {
        private psjldb12Context db = new psjldb12Context();
```

```csharp
private AppService apps = new AppService();
//
//GET:/GroupList/
public virtual ActionResult Index(FormCollection fc)
{
    string gid = " ", gname = " ";
    int rps = 20;
    if(fc.AllKeys.Count() > 0)
    {
        apps.SetSession("grid_groupid", fc["grid_groupid"]);
        apps.SetSession("grid_groupname", fc["grid_groupname"]);
        apps.SetSession("grid_rowsperpage", fc["grid_rowsperpage"]);
    }
    gid = apps.GetSession("grid_groupid");
    gname = apps.GetSession("grid_groupname");
    rps = String.IsNullOrEmpty(apps.GetSession("grid_rowsperpage")) ? rps : Convert.ToInt32(apps.GetSession("grid_rowsperpage"));
    ViewBag.groupid = gid;
    ViewBag.groupname = gname;
    ViewBag.rowsperpage = rps;
    var rd = db.KGroupLists.Where(g => g.groupid.Contains(gid) && g.groupname.Contains(gname)).OrderBy(g => g.groupid);
    return View(rd.ToList());
}
//
//GET:/GroupList/Details/5
public virtual ActionResult Details(string id = null)
{
    KGroupList grouplist = db.KGroupLists.Find(id);
    if(grouplist == null)
    {
        return HttpNotFound();
    }
    return View(grouplist);
}
/// <summary>
///远程校验角色编号方法:CodeValidate
/// </summary>
/// <param name = "groupid"> </param>
/// <returns> bool </returns>
```

```csharp
[HttpGet]
public virtual ActionResult CodeValidate(string groupid = null)
{
    var rd = db.KGroupLists.Find(groupid.ToUpper());
    bool exists = true;
    if(rd == null)
    {
        exists = false;
    }
    return Json(!exists, JsonRequestBehavior.AllowGet);
}
//
//GET:/GroupList/Create
public virtual ActionResult Create()
{
    ViewBag.Message = "根据提示输入数据...";
    return View();
}
//
//POST:/GroupList/Create
[HttpPost]
public virtual ActionResult Create(KGroupList grouplist)
{
    if(ModelState.IsValid)
    {
        grouplist.groupid.ToUpper();
        db.KGroupLists.Add(grouplist);
        db.SaveChanges();
        ViewBag.Message = "数据存储完成...";
        //return RedirectToAction("Index");
    }
    else
    {
        ViewBag.Message = "数据有错误,存储失败...";
    }
    return View(grouplist);
}
//
//GET:/GroupList/Edit/5
public virtual ActionResult Edit(string id = null)
```

```csharp
{
    KGroupList grouplist = db.KGroupLists.Find(id);
    if(grouplist == null)
    {
        return HttpNotFound();
    }
    return View(grouplist);
}
//
//POST:/GroupList/Edit/5
[HttpPost]
public virtual ActionResult Edit(KGroupList grouplist)
{
    if(ModelState.IsValid)
    {
        db.Entry(grouplist).State = EntityState.Modified;
        db.SaveChanges();
        ViewBag.Message = "数据存储成功...";
        //return RedirectToAction("Index");
    }
    else
    {
        ViewBag.Message = "数据输入有错误!";
    }
    return View(grouplist);
}
//
//GET:/GroupList/Delete/5
public virtual ActionResult Delete(string id = null)
{
    KGroupList grouplist = db.KGroupLists.Find(id);
    if(grouplist == null)
    {
        return HttpNotFound();
    }
    return View(grouplist);
}
//
//POST:/GroupList/Delete/5
[HttpPost,ActionName("Delete")]
```

```
            public virtual ActionResult DeleteConfirmed(string id)
            {
                KGroupList grouplist = db.KGroupLists.Find(id);
                db.KGroupLists.Remove(grouplist);
                db.SaveChanges();
                return RedirectToAction("Index");
            }
            protected override void Dispose(bool disposing)
            {
                db.Dispose();
                base.Dispose(disposing);
            }
        }
    }
```

控制器的组成内容分为以下几个部分：

（1）using 语句引入所需要的类库；

（2）定义命名空间"namespace psjlmvc4.Areas.KArea.Controllers"，即控制器的存储目录；

（3）定义类控制器"KGroupListController"的实例变量 db 和 apps，以实现数据库管理操作和公用方法调用；

（4）以"ActionResult"为返回类型的方法。

6.1.4 记录数据检索方法

CRUD 控制器中的数据检索方法名称是"Index"，主要功能是根据参数（可有可无），从数据库中检索相关数据记录，并填充对应的实体模型，然后将实体模型传递给视图显示数据。其原始代码内容如下：

```
public virtual ActionResult Index()
{
    return View(db.KGroupLists.ToList());
}
```

默认生成的方法中的操作逻辑是检索所有的数据记录。本例中的代码内容是经开发者修改扩展后的代码内容，不是由模板生成出来的。

6.1.5 记录详细内容显示方法

数据记录详细内容显示的方法名称是"Details"，根据"Index"视图中的

"详细（Details）"链接传递的记录 ID，检索相应的记录，然后传递给相应视图，实现记录内容的显示，其原始代码内容如下：

```
//GET:/GroupList/Details/5
public virtual ActionResult Details(string id = null)
{
    KGroupList grouplist = db.KGroupLists.Find(id);
    if(grouplist = = null)
    {
        return HttpNotFound();
    }
    return View(grouplist);
}
```

6.1.6 新增记录方法

新增记录方法的名称是"Create"，在控制器有两种存在形态，一种是无参数形态，一种是有模型参数形态。前者是数据编辑视图的入口，后者是数据编辑完成后实现数据存储功能，二者的原始代码内容如下：

```
//GET:/GroupList/Create
public virtual ActionResult Create()
{
    ViewBag.Message ="根据提示输入数据...";
    return View();
}
//POST:/GroupList/Create
[HttpPost]
public virtual ActionResult Create(KGroupList grouplist)
{
    if(ModelState.IsValid)
    {
        db.KGroupLists.Add(grouplist);
        db.SaveChanges();
        ViewBag.Message ="数据存储完成...";
        //return RedirectToAction("Index");
    }
    else
```

```
        {
            ViewBag.Message = "数据有错误,存储失败...";
        }
    return View(grouplist);
}
```

6.1.7 记录数据编辑方法

记录数据编辑方法的名称是"Edit",在控制器中存在两种形态,一种是编辑视图的入口方法,另一种是编辑完成后实现数据存储的方法,前者接受参数是记录ID,后者接受参数是记录ID和记录数据,其原始代码内容如下:

```
//GET:/GroupList/Edit/5
public virtual ActionResult Edit(string id = null)
{
    KGroupList grouplist = db.KGroupLists.Find(id);
    if(grouplist == null)
    {
        return HttpNotFound();
    }
    return View(grouplist);
}
//POST:/GroupList/Edit/5
[HttpPost]
public virtual ActionResult Edit(KGroupList grouplist)
{
    if(ModelState.IsValid)
    {
        db.Entry(grouplist).State = EntityState.Modified;
        db.SaveChanges();
        ViewBag.Message = "数据存储成功...";
        //return RedirectToAction("Index");
    }
    else
    {
        ViewBag.Message = "数据输入有错误!";
    }
    return View(grouplist);
}
```

6.1.8 记录删除方法

记录删除方法的名称是"Delete",在控制器中存在两种形态,一种是删除确认视图方法,另一种是确认后完成删除功能的方法,其原始代码内容如下:

```
//GET:/GroupList/Delete/5
public virtual ActionResult Delete(string id = null)
{
    KGroupList grouplist = db.KGroupLists.Find(id);
    if(grouplist = = null)
    {
        return HttpNotFound();
    }
    return View(grouplist);
}
//POST:/GroupList/Delete/5
[HttpPost,ActionName("Delete")]
public virtual ActionResult DeleteConfirmed(string id)
{
    KGroupList grouplist = db.KGroupLists.Find(id);
    db.KGroupLists.Remove(grouplist);
    db.SaveChanges();
    return RedirectToAction("Index");
}
```

注意:"Delete"方法的两种形态的名称并不一致,不过通过实体域属性"[HttpPost,ActionName("Delete")]"定义后,将两者联系在一起,其作用是将两者视同为名称一样的一个方法的两种形态。

6.2 CRUD 视图模板应用实例

在"添加基架"对话框选择"包含视图的 MVC 5 控制器(使用 Entity Framework)"项目,在生成 CRUD 控制器的同时,同样根据视图模板也生成了实际使用的 CRUD 方法对应的 CRUD 视图,并分别以名称"Index"、"Create"、"Edit"、"Delete"、"Details"进行命名,存储于"Views/KGrouplist/"目录内,如图 6-4 所示。

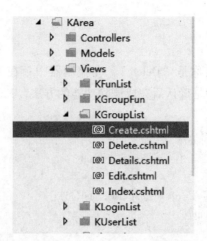

图 6-4　Views/KGrouplist 中的视图文件

6.2.1　记录列表显示视图

由模板生成的记录显示视图名称是"Index. cshtml"，由控制器中的"Index"方法调用，其代码内容如下：

```
@ model IEnumerable < psjlmvc4. Models. KGroupList >
@ {
    ViewBag. Title = "KGroupList 记录数据列表(Index For Record Data)";
}
< table id = "dtable" >
    < caption > < h2 > @ ViewBag. Title < /h2 > < /caption >
    < thead >
    < tr >
        < th >
            @ Html. DisplayNameFor( model = > model. groupid)
        < /th >
        < th >
            @ Html. DisplayNameFor( model = > model. groupname)
        < /th >
        < th >
            @ Html. DisplayNameFor( model = > model. remark)
        < /th >
        < th > < /th >
    < /tr >
    < /thead > < tbody >
```

```
@foreach(var item in Model){
    <tr>
        <td>
            @Html.DisplayFor(modelItem => item.groupid)
        </td>
        <td>
            @Html.DisplayFor(modelItem => item.groupname)
        </td>
        <td>
            @Html.DisplayFor(modelItem => item.remark)
        </td>
        <td>
            @Html.ActionLink("编辑","Edit",new{id=item.statusid})|
            @Html.ActionLink("详细","Details",new{id=item.statusid})|
            @Html.ActionLink("删除","Delete",new{id=item.statusid})
        </td>
    </tr>
}
    <tbody>
</table>
<p>
    @Html.ActionLink("新增","Create",null,new{@class="alinkcss"})
    @Html.ActionLink("打印","Print",null,new{@class="alinkcss"})
    @Html.ActionLink("返回","Index","Home",new{Area=""},new{@class="alinkcss"})
</p>
```

其组成内容说明如下：

（1）以"@model"语句开始，接受由方法传递过来的列表类型的记录数据，转换为视图数据"Model"；

（2）以表格（Table）形式显示模型中的记录数据；

（3）每个记录附加"编辑"、"详细"和"删除"三个链接，以实现相应的功能；

（4）增加新记录的链接"新增"，连同"返回"链接放在表格的后边。

其运行效果如图6-5所示。

由于实际的角色实体显示视图（Index.cshtml）在此基础上做了大量的内容修改，已经不是CRUD模板生成的原始内容，因此，在此借用了工程状态实体的显示视图内容，从而说明CRUD视图模板所生成的原始显示视图内容。

工程状态编号	工程状态名称	工程状态说明	
1	备验	上海市呈现叶落归根埕 吉一个中心	编辑 | 详细 | 删除
2	停工	无奇不有	编辑 | 详细 | 删除
3	完工	地球磁场	编辑 | 详细 | 删除
4	在监	天干压垮	编辑 | 详细 | 删除
5	暂缓	无奇不有地球磁场	编辑 | 详细 | 删除
6	待审	工程目前停工等待上级部门审查	编辑 | 详细 | 删除
7	不确定	博士十一下雨	编辑 | 详细 | 删除
8	新状态	王一个中心	编辑 | 详细 | 删除

新增　打印　返回

图 6-5　Index 视图运行后的效果图

6.2.2　记录新增显示视图

由模板生成的记录新增视图名称是"Create.cshtml",在控制器中对应同名"Create"的两个方法为：一个是无参数调用、进入编辑视图的入口功能；另一个是通过"HttpPost"方式,接受模型参数(记录内容输入后提交)实现记录存储的功能,其代码内容如下：

```
@model psjlmvc4.Models.KGroupList
@{
    ViewBag.Title = "新增系统用户角色";
}
@using(Html.BeginForm()){
    @Html.ValidationSummary(true)
    <fieldset>
        <legend>@ViewBag.Title</legend>
        <div class="editor-label">
            @Html.LabelFor(model => model.groupid)
            @Html.EditorFor(model => model.groupid)
            @Html.ValidationMessageFor(model => model.groupid)
        </div>
        <div class="editor-label">
            @Html.LabelFor(model => model.groupname)
            @Html.EditorFor(model => model.groupname)
            @Html.ValidationMessageFor(model => model.groupname)
        </div>
        <div class="editor-label">
            @Html.LabelFor(model => model.remark)
            @Html.EditorFor(model => model.remark)
```

```
            @Html.ValidationMessageFor(model => model.remark)
        </div>
        <hr/>
        <p>
            <button type="submit" class="btn btn-primary glyphicon glyphicon-save">确认新增</button>
            @Html.ActionLink("返回列表","Index",null,new{ @class = "btn btn-primary glyphicon glyphicon-list"})
        </p>
        <hr/>
        <div>操作提示：</div>
        <ol>
            <li>系统角色编号是一位大写英文字母,例如,"A"--"系统管理"角色;</li>
            <li>系统角色名称力求简洁明确,含义准确,长度不能超过20个字;</li>
            <li>角色编号一经定义存储,不能修改;</li>
            <li>通过删除某个系统角色,再新增,完成编号修改。</li>
        </ol>
        <hr/>
        <span>@ViewBag.Message</span>
    </fieldset>
}
@section Scripts{
    @Scripts.Render("~/bundles/jqueryval")
}
```

其中,"<ol>……</ol>"是后期新增内容。视图执行后的效果如图6-6所示。

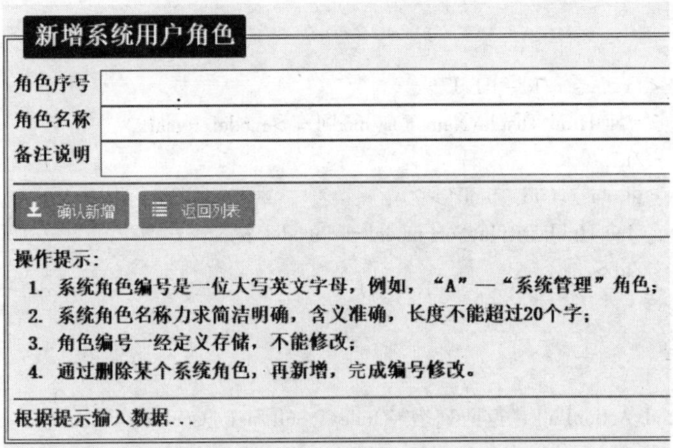

图6-6 Create 视图运行后的效果图

6.2.3 记录详细内容显示视图

由模板生成的记录详细内容显示视图名称是"Details.cshtml",对应控制器中的方法是"Details"方法调用,其代码内容如下:

```
@model psjlmvc4.Models.KGroupList
@{
    ViewBag.Title = "系统角色记录详细内容";
}
<table id="tt-table">
    <caption>@ViewBag.Title</caption>
    <tr>
        <td class="td-label">
            @Html.DisplayNameFor(model => model.groupid)
        </td>
        <td class="td-field">
            @Html.DisplayFor(model => model.groupid)
        </td>
    </tr>
    <tr>
        <td class="td-label">
            @Html.DisplayNameFor(model => model.groupname)
        </td>
        <td class="td-field">
            @Html.DisplayFor(model => model.groupname)
        </td>
    </tr>
    <tr>
        <td class="td-label">
            @Html.DisplayNameFor(model => model.remark)
        </td>
        <td class="td-field">
            @Html.DisplayFor(model => model.remark)
        </td>
    </tr>
</table>
<p>
    @Html.ActionLink("返回列表","Index",null,new{@class="btn btn-primary glyphicon glyphicon-list"})
</p>
```

视图执行后的效果如图 6-7 所示。

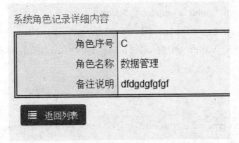

图 6-7 Details 视图运行后的效果图

6.2.4 记录编辑显示视图

由模板生成的记录编辑显示视图名称是"Edit.cshtml",在控制器中对应同名"Edit"的两个方法为：一个是带记录主键参数调用、进入编辑视图的入口方法；另一个是通过"HttpPost"方式，接受模型参数（记录内容输入后提交）实现记录存储的方法，其代码内容如下：

```
@ model psjlmvc4.Models.KGroupList
@ {
    ViewBag.Title = "系统角色编辑";
}
@ using( Html.BeginForm( ) )
{
    @ Html.ValidationSummary( true )
    <fieldset>
        <legend> @ ViewBag.Title </legend>
        <div>
            @ Html.LabelFor( model = > model.groupid )
            @ Html.DisplayFor( model = > model.groupid )
            @ Html.HiddenFor( model = > model.groupid )
        </div>
        <hr/>
        <div class = "editor - label" >
            @ Html.LabelFor( model = > model.groupname )
            @ Html.EditorFor( model = > model.groupname )
            @ Html.ValidationMessageFor( model = > model.groupname )
        </div>
```

```
            <div class = "editor - label">
                @Html.LabelFor(model = > model.remark)
                @Html.EditorFor(model = > model.remark)
                @Html.ValidationMessageFor(model = > model.remark)
            </div>
            <hr/>
            <p>
                <button type = "submit"class = "btn btn - primary glyphicon glyphicon - save">确认存储</button>
                @Html.ActionLink("返回列表","Index",null,new{@class = "btn btn - primary glyphicon glyphicon - list"})
            </p>
            <span>@ViewBag.ErrorMessage</span>
        </fieldset>
}
@section Scripts{
    @Scripts.Render("~/bundles/jqueryval")
}
```

运行后的效果如图 6-8 所示。

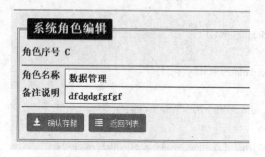

图 6-8 Edit 视图运行后的效果图

6.2.5 记录删除显示视图

由模板生成的记录编辑显示视图名称是"Delete.cshtml",和"Edit"方法一样,在控制器中对应同名"Delete"的两个方法为:一个是带记录主键参数调用、进入记录内容显示视图的入口方法;另一个是通过"HttpPost"方式,接受记录主键参数(确认后提交)实现记录删除和存储的方法,其代码内容如下:

```
@model psjlmvc4.Models.KGroupList
@{
    ViewBag.Title = "系统用户角色删除";
}
<h2>确认删除此记录吗？</h2>
<fieldset>
    <legend>@ViewBag.Title</legend>
    <div class="display-label">
        @Html.DisplayNameFor(model => model.groupid)
        @Html.DisplayFor(model => model.groupid)
    </div>
    <div class="display-label">
        @Html.DisplayNameFor(model => model.groupname)
        @Html.DisplayFor(model => model.groupname)
    </div>
    <div class="display-label">
        @Html.DisplayNameFor(model => model.remark)
        @Html.DisplayFor(model => model.remark)
    </div>
</fieldset>
@using(Html.BeginForm()){
    <p>
        <button type="submit" class="btn btn-primary glyphicon glyphicon-save">确认删除</button>
        @Html.ActionLink("返回列表","Index",null,new{@class="btn btn-primary glyphicon glyphicon-list"})
    </p>
}
```

运行后的视图效果如图6-9所示。

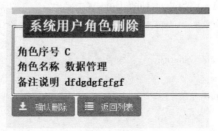

图6-9 Delete视图运行后的效果图

上述由模板生成的有关模型记录操作 CRUD 视图是根据修改后的模板生成的，有关系统提供的模板和如何根据自己的需要进行个性化设置会在以下章节中说明。

6.3 ASP.NET MVC 系统自有 CRUD 模板

在 MVC 框架下进行日常的应用程序开发中，为了减少开发人员写代码的时间，基于 ASP.NET MVC 的 VS2013 工具提供了生成 CRUD 操作的脚手架模板，包括模型控制器和 5 个交互显示视图模板。下面以包含视图生成模板的 MVC 4 控制器（使用 Entity Framework）模板为例，说明系统提供的模板内容及功能。

6.3.1 控制器生成模板

在 MVC4 框架中，系统提供的控制器生成模板文件存储于目录"C：\Program Files\Microsoft Visual Studio 12.0\Common7\IDE\ItemTemplates\CSharp\Web\MVC 4\CodeTemplates\AddController"中，其中模板文件"ControllerWithContext.tt"是用于生成以 C#语言为引擎的控制器的自动生成模板。以下是"ControllerWithContext.tt"的内容：

```
<#@ template language="C#"HostSpecific="True"#>
<#
var Model = (MvcTextTemplateHost)Host;
var routePrefix = String.Empty;
if(! String.IsNullOrWhiteSpace(Model.AreaName)){
    routePrefix = "/" + Model.AreaName;
}
routePrefix += "/" + Model.ControllerRootName + "/";
#>
<#@ import namespace="System.Collections"#>
<#@ import namespace="System.Collections.Generic"#>
<#@ import namespace="System.Data.Objects"#>
<#@ import namespace="System.Linq"#>
using System;
using System.Collections.Generic;
using System.Data;
using System.Data.Entity;
using System.Linq;
using System.Web;
using System.Web.Mvc;
```

```
<# if ( MvcTextTemplateHost. NamespaceNeeded ( Model. Namespace, Model. ModelType.
Namespace)){#>
    using <# = Model. ModelType. Namespace#>;
    <#}#>
    <# if ( Model. ContextType. Namespace! = Model. ModelType. Namespace && MvcTextTem-
plateHost. NamespaceNeeded( Model. Namespace, Model. ContextType. Namespace)){#>
    using <# = Model. ContextType. Namespace#>;
    <#}#>
    namespace <# = Model. Namespace#>
    {
    <#
        var modelName = Model. ModelType. Name;
        var entitySetName = Model. EntitySetName;
        var modelVariable = modelName. ToLowerInvariant();
        var entitySetVariable = entitySetName. ToLowerInvariant();
        var primaryKey = Model. PrimaryKeys[0];
        var lambdaVar = modelVariable[0];
        var isObjectContext = typeof( ObjectContext). IsAssignableFrom( Model. ContextType);
    #>
        public class <# = Model. ControllerName #>:Controller
        {
            private <# = Model. ContextType. Name #> db = new <# = Model. ContextType. Name
#>();
            //
            //GET: <# = routePrefix #>
            public ActionResult Index()
            {
    <#  var includeExpressions ="";
        if( isObjectContext){
            includeExpressions = String. Join(" ", Model. RelatedProperties. Values. Select( prop-
erty = > String. Format(". Include( \"{0}\")", property. PropertyName)));
        }
        else {
            includeExpressions = String. Join(" ", Model. RelatedProperties. Values. Select( prop-
erty = > String. Format(". Include({0} = >{0}.{1})", lambdaVar, property. PropertyName)));
        }
    #>
        <# if(! String. IsNullOrEmpty( includeExpressions)){#>
                var <# = entitySetVariable#> = db. <# = entitySetName #> <# = includeEx-
pressions #>;
```

```
                return View( <#= entitySetVariable #>.ToList( ) );
    <#} else {#>
                return View( db.<#= entitySetName #> <#= includeExpressions #>.ToList
( ) );
    <#}#>
            }
            //
            //GET: <#= routePrefix #>Details/5
            public ActionResult Details( <#= primaryKey.ShortTypeName #> id = <#= prima-
ryKey.DefaultValue #> )
            {
    <# if( isObjectContext ) {#>
                <#= modelName #> <#= modelVariable #> = db.<#= entitySetName #>.
Single( <#= lambdaVar #> => <#= lambdaVar #>.<#= primaryKey.Name #> == id );
    <#} else {#>
                <#= modelName #> <#= modelVariable #> = db.<#= entitySetName #>.
Find( id );
    <#}#>
                if( <#= modelVariable #> == null )
                {
                    return HttpNotFound( );
                }
                return View( <#= modelVariable #> );
            }
            //
            //GET: <#= routePrefix #>Create
            public ActionResult Create( )
            {
    <# foreach( var property in Model.RelatedProperties.Values ) {#>
                ViewBag.<#= property.ForeignKeyPropertyName #> = new SelectList( db.<
#= property.EntitySetName #>, " <#= property.PrimaryKey #>"," <#= proper-
ty.DisplayPropertyName #>" );
    <# } #>
                return View( );
            }
            //
            //POST: <#= routePrefix #>Create
            [HttpPost]
            [ValidateAntiForgeryToken]
            public ActionResult Create( <#= modelName #> <#= modelVariable #> )
```

```
                {
                    if(ModelState.IsValid)
                    {
<# if(primaryKey.Type == typeof(Guid)){#>
                        <#= modelVariable #>.<#= primaryKey.Name #> = Guid.NewGuid();
<#}#>
<# if(isObjectContext){#>
                        db.<#= entitySetName #>.AddObject(<#= modelVariable #>);
<#} else {#>
                        db.<#= entitySetName #>.Add(<#= modelVariable #>);
<#}#>
                        db.SaveChanges();
                        return RedirectToAction("Index");
                    }
<# foreach(var property in Model.RelatedProperties.Values){#>
                    ViewBag.<#= property.ForeignKeyPropertyName #> = new SelectList(db.<#= property.EntitySetName #>,"<#= property.PrimaryKey #>","<#= property.DisplayPropertyName #>",<#= modelVariable #>.<#= property.ForeignKeyPropertyName #>);
<# } #>
                    return View(<#= modelVariable #>);
                }

                //
                //GET: <#= routePrefix #>Edit/5
                public ActionResult Edit(<#= primaryKey.ShortTypeName #> id = <#= primaryKey.DefaultValue #>)
                {
<# if(isObjectContext){#>
                    <#= modelName #> <#= modelVariable #> = db.<#= entitySetName #>.Single(<#= lambdaVar #> => <#= lambdaVar #>.<#= primaryKey.Name #> == id);
<#} else {#>
                    <#= modelName #> <#= modelVariable #> = db.<#= entitySetName #>.Find(id);
<#}#>
                    if(<#= modelVariable #> == null)
                    {
                        return HttpNotFound();
                    }
<# foreach(var property in Model.RelatedProperties.Values){#>
```

```
                    ViewBag. <# = property.ForeignKeyPropertyName #> = new SelectList(db.<
# = property.EntitySetName #>,"<# = property.PrimaryKey #>","<# = proper-
ty.DisplayPropertyName #>", <# = modelVariable #>. <# = property.ForeignKeyPropertyName #
>);
    <#}#>
                    return View(<# = modelVariable #>);
            }
        //
        //POST: <# = routePrefix #>Edit/5
        [HttpPost]
        [ValidateAntiForgeryToken]
        public ActionResult Edit(<# = modelName #> <# = modelVariable #>)
        {
                if(ModelState.IsValid)
                {
    <# if(isObjectContext){#>
                    db.<# = entitySetName #>.Attach(<# = modelVariable #>);
                    db.ObjectStateManager.ChangeObjectState(<# = modelVariable #>,En-
tityState.Modified);
    <#} else {#>
                    db.Entry(<# = modelVariable #>).State = EntityState.Modified;
    <#}#>
                    db.SaveChanges();
                    return RedirectToAction("Index");
                }
    <# foreach(var property in Model.RelatedProperties.Values){#>
                    ViewBag.<# = property.ForeignKeyPropertyName #> = new SelectList(db.<
# = property.EntitySetName #>,"<# = property.PrimaryKey #>","<# = proper-
ty.DisplayPropertyName #>", <# = modelVariable #>. <# = property.ForeignKeyPropertyName #
>);
    <#}#>
                    return View(<# = modelVariable #>);
            }
        //
        //GET: <# = routePrefix #>Delete/5
        public ActionResult Delete(<# = primaryKey.ShortTypeName #> id = <# = prima-
ryKey.DefaultValue #>)
            {
    <# if(isObjectContext){#>
```

```
                <#= modelName #> <#= modelVariable #> = db.<#= entitySetName #>.
Single( <#= lambdaVar #> => <#= lambdaVar #>.<#= primaryKey.Name #> == id);
    <#} else {#>
                <#= modelName #> <#= modelVariable #> = db.<#= entitySetName #>.
Find(id);
    <#}#>
                if( <#= modelVariable #> == null)
                {
                    return HttpNotFound();
                }
                return View( <#= modelVariable #> );
            }
            //
            //POST: <#= routePrefix #>Delete/5
            [HttpPost,ActionName("Delete")]
            [ValidateAntiForgeryToken]
            public ActionResult DeleteConfirmed( <#= primaryKey.ShortTypeName #> id)
            {
<# if( isObjectContext) {#>
                <#= modelName #> <#= modelVariable #> = db.<#= entitySetName #>.
Single( <#= lambdaVar #> => <#= lambdaVar #>.<#= primaryKey.Name #> == id);
                db.<#= entitySetName #>.DeleteObject( <#= modelVariable #> );
    <#} else {#>
                <#= modelName #> <#= modelVariable #> = db.<#= entitySetName #>.
Find(id);
                db.<#= entitySetName #>.Remove( <#= modelVariable #> );
    <#}#>
                db.SaveChanges();
                return RedirectToAction("Index");
            }
            protected override void Dispose(bool disposing)
            {
                db.Dispose();
                base.Dispose(disposing);
            }
        }
    }
```

　　这就是用于自动生成基于模型的 CRUD 操作的控制器模板，ASP.NET MVC 称之为"T4"或"TT"模板。从功能上划分，文件内容可以分为以下三个部分：

（1）以"<#@ …… #>"为标记的语句行，其中内容为命令，用于提供控制模板的处理方法和动态生成的内容。

（2）以"<# …… #>"为标记的语句块，为代码执行控制块，根据不同的情况，选择生成相应的内容，例如：<##>——代码表达式；<#=#>——显示表达式值；<#+#>——声明定义方法和变量。

（3）没有标记的内容，即静态内容，为模板直接输入内容，也就是说，这部分内容是原样复制到生成的控制器的内容。

在利用模板生成控制器时，模板的任务有：确定编程语言和生成的控制器文件类别（CS 或 VB）；引入必要的类库；分别定义 CRUD 操作实现的模板格式。

6.3.2　记录列表显示视图生成模板

系统提供了六个视图生成模板：四个 CURD 视图，一个记录详细内容显示视图和一个没有模型的空视图，相应的视图模板文件位于"C：\Program Files\Microsoft Visual Studio 12.0\Common7\IDE\ItemTemplates\CSharp\Web\MVC 4\CodeTemplates\AddView\CSHTML"，其中文件如下（以 VB 为语言的模板在此省略）：List.tt——记录列表显示视图模板；Create.tt——新增记录视图模板；Details.tt——记录详细内容显示视图模板；Edit.tt——编辑记录视图模板；Delete.tt——删除记录视图模板；Empty.tt——空视图模板。

记录列表显示视图生成模板的内容如下：

```
<#@ template language = "C#"HostSpecific = "True"#>
<#@ output extension = ".cshtml"#>
<#@ assembly name = "System.ComponentModel.DataAnnotations"#>
<#@ assembly name = "System.Core"#>
<#@ assembly name = "System.Data.Entity"#>
<#@ assembly name = "System.Data.Linq"#>
<#@ import namespace = "System"#>
<#@ import namespace = "System.Collections.Generic"#>
<#@ import namespace = "System.ComponentModel.DataAnnotations"#>
<#@ import namespace = "System.Data.Linq.Mapping"#>
<#@ import namespace = "System.Data.Objects.DataClasses"#>
<#@ import namespace = "System.Linq"#>
<#@ import namespace = "System.Reflection"#>
<#@ import namespace = "Microsoft.VisualStudio.Web.Mvc.Scaffolding.BuiltIn"#>
<#
MvcTextTemplateHost mvcHost = MvcTemplateHost;
#>
@model IEnumerable<#="<" + mvcHost.ViewDataTypeName + ">"#>
```

```
<#
//The following chained if - statement outputs the file header code and markup for a partial view, a content page, or a regular view.
if( mvcHost. IsPartialView) {
#>
<#
} else if( mvcHost. IsContentPage) {
#>
@{
    ViewBag. Title = "<# = mvcHost. ViewName#>";
<#
if( ! String. IsNullOrEmpty( mvcHost. MasterPageFile) ) {
#>
    Layout = "<# = mvcHost. MasterPageFile#>";
<#
}
#>
}
<h2><# = mvcHost. ViewName#></h2>
<#
} else {
#>
@{
    Layout = null;
}
<! DOCTYPE html>
<html>
<head>
    <meta name = "viewport"content = "width = device - width"/>
    <title><# = mvcHost. ViewName #></title>
</head>
<body>
<#
    PushIndent("    ");
}
#>
<p>
    @ Html. ActionLink("Create New","Create")
</p>
<table>
```

```
        <tr>
<#
List<ModelProperty> properties = GetModelProperties(mvcHost.ViewDataType);
foreach(ModelProperty property in properties){
    if(!property.IsPrimaryKey && property.Scaffold){
#>
            <th>
                @Html.DisplayNameFor(model => model.<#=property.ValueExpression#>)
            </th>
<#
    }
}
#>
            <th></th>
        </tr>
@foreach(var item in Model){
        <tr>
<#
foreach(ModelProperty property in properties){
    if(!property.IsPrimaryKey && property.Scaffold){
#>
            <td>
                @Html.DisplayFor(modelItem => <#=property.ItemValueExpression#>)
            </td>
<#
    }
}
string pkName = GetPrimaryKeyName(mvcHost.ViewDataType);
if(pkName != null){
#>
            <td>
                @Html.ActionLink("Edit","Edit",new{id = item.<#=pkName#>}) |
                @Html.ActionLink("Details","Details",new{id = item.<#=pkName#>}) |
                @Html.ActionLink("Delete","Delete",new{id = item.<#=pkName#>})
            </td>
<#
} else {
#>
            <td>
```

```
                @Html.ActionLink("Edit","Edit",new{/* id=item.PrimaryKey */})|
                @Html.ActionLink("Details","Details",new{/* id=item.PrimaryKey */})|
                @Html.ActionLink("Delete","Delete",new{/* id=item.PrimaryKey */})
            </td>
<#
}
#>
        </tr>
}
</table>
<#
//The following code closes the asp:Content tag used in the case of a master page and the body and html tags in the case of a regular view page
#>
<#
if(mvcHost.IsContentPage){
#>
<#
}else if(! mvcHost.IsPartialView && ! mvcHost.IsContentPage){
    ClearIndent();
#>
</body>
</html>
<#
}
#>
<#+
//Describes the information about a property on the model
class ModelProperty{
    public string Name{get;set;}
    public string AssociationName{get;set;}
    public string ValueExpression{get;set;}
    public string ModelValueExpression{get;set;}
    public string ItemValueExpression{get;set;}
    public Type UnderlyingType{get;set;}
    public bool IsPrimaryKey{get;set;}
    public bool IsForeignKey{get;set;}
    public bool IsReadOnly{get;set;}
    public bool Scaffold{get;set;}
```

```csharp
    }
    //Change this list to include any non-primitive types you think should be eligible for display/edit
    static Type[] bindableNonPrimitiveTypes = new[]{
        typeof(string),
        typeof(decimal),
        typeof(Guid),
        typeof(DateTime),
        typeof(DateTimeOffset),
        typeof(TimeSpan),
    };

    //Call this to get the list of properties in the model. Change this to modify or add your
    //own default formatting for display values.
    List<ModelProperty> GetModelProperties(Type type){
        List<ModelProperty> results = GetEligibleProperties(type);
        foreach(ModelProperty prop in results){
            if(prop.UnderlyingType == typeof(double) || prop.UnderlyingType == typeof(decimal)){
                prop.ModelValueExpression = " String.Format ( \"{0: F} \"," + prop.ModelValueExpression + ")";
            }
            else if(prop.UnderlyingType == typeof(DateTime)){
                prop.ModelValueExpression = " String.Format ( \"{0: g} \"," + prop.ModelValueExpression + ")";
            }
        }
        return results;
    }
    //Call this to determine if property has scaffolding enabled
    bool Scaffold(PropertyInfo property){
        foreach(object attribute in property.GetCustomAttributes(true)){
            var scaffoldColumn = attribute as ScaffoldColumnAttribute;
            if(scaffoldColumn != null && ! scaffoldColumn.Scaffold){
                return false;
            }
        }
        return true;
    }
```

```csharp
//Call this to determine if the property represents a primary key. Change the
//code to change the definition of primary key.
bool IsPrimaryKey(PropertyInfo property){
        if(string.Equals(property.Name,"id",StringComparison.OrdinalIgnoreCase)){//EF Code First convention
            return true;
        }
        if(string.Equals(property.Name, property.DeclaringType.Name + "id",StringComparison.OrdinalIgnoreCase)){//EF Code First convention
            return true;
        }
        foreach(object attribute in property.GetCustomAttributes(true)){
            if(attribute is KeyAttribute){//WCF RIA Services and EF Code First explicit
                return true;
            }
            var edmScalar = attribute as EdmScalarPropertyAttribute;
            if(edmScalar! = null && edmScalar.EntityKeyProperty){//EF traditional
                return true;
            }
            var column = attribute as ColumnAttribute;
            if(column! = null && column.IsPrimaryKey){//LINQ to SQL
                return true;
            }
        }
        return false;
}
//This will return the primary key property name,if and only if there is exactly
//one primary key. Returns null if there is no PK,or the PK is composite.
string GetPrimaryKeyName(Type type){
    IEnumerable<string> pkNames = GetPrimaryKeyNames(type);
    return pkNames.Count() = =1? pkNames.First():null;
}
//This will return all the primary key names. Will return an empty list if there are none.
IEnumerable<string> GetPrimaryKeyNames(Type type){
    return GetEligibleProperties(type).Where(mp = > mp.IsPrimaryKey).Select(mp = > mp.Name);
}
//Call this to determine if the property represents a foreign key.
bool IsForeignKey(PropertyInfo property){
    return MvcTemplateHost.RelatedProperties.ContainsKey(property.Name);
```

```
}
//A foreign key, e. g. CategoryID, will have a value expression of Category. CategoryID
string GetValueExpressionSuffix( PropertyInfo property) {
    RelatedModel propertyModel;
    MvcTemplateHost. RelatedProperties. TryGetValue( property. Name, out propertyModel);
    return propertyModel! = null? propertyModel. PropertyName +"." + propertyModel. DisplayPropertyName: property. Name;
}
//A foreign key, e. g. CategoryID, will have an association name of Category
string GetAssociationName( PropertyInfo property) {
    RelatedModel propertyModel;
    MvcTemplateHost. RelatedProperties. TryGetValue( property. Name, out propertyModel);
    return propertyModel! = null? propertyModel. PropertyName: property. Name;
}
//Helper
List < ModelProperty > GetEligibleProperties( Type type) {
    List < ModelProperty > results = new List < ModelProperty > ();
    foreach( PropertyInfo prop in type. GetProperties( BindingFlags. Public | BindingFlags. Instance)) {
        Type underlyingType = Nullable. GetUnderlyingType( prop. PropertyType) ?? prop. PropertyType;
        if( prop. GetGetMethod()! = null && prop. GetIndexParameters(). Length = = 0
            && IsBindableType( underlyingType)) {
            string valueExpression = GetValueExpressionSuffix( prop);
            results. Add( new ModelProperty {
                Name = prop. Name,
                AssociationName = GetAssociationName( prop),
                ValueExpression = valueExpression,
                ModelValueExpression = "Model. " + valueExpression,
                ItemValueExpression = "item. " + valueExpression,
                UnderlyingType = underlyingType,
                IsPrimaryKey = IsPrimaryKey( prop),
                IsForeignKey = IsForeignKey( prop),
                IsReadOnly = prop. GetSetMethod() = = null,
                Scaffold = Scaffold( prop)
            });
        }
    }
    return results;
}
```

```
//Helper
bool IsBindableType(Type type){
    return type.IsPrimitive||bindableNonPrimitiveTypes.Contains(type);
}
MvcTextTemplateHost MvcTemplateHost{
    get{
        return(MvcTextTemplateHost)Host;
    }
}
#>
```

6.3.3 新增记录显示视图生成模板

新增记录显示视图生成模板"Create.tt"的内容如下:

```
<#@ template language="C#"HostSpecific="True"#>
<#@ assembly name="System.ComponentModel.DataAnnotations"#>
<#@ assembly name="System.Core"#>
<#@ assembly name="System.Data.Entity"#>
<#@ assembly name="System.Data.Linq"#>
<#@ import namespace="System"#>
<#@ import namespace="System.Collections.Generic"#>
<#@ import namespace="System.ComponentModel.DataAnnotations"#>
<#@ import namespace="System.Data.Linq.Mapping"#>
<#@ import namespace="System.Data.Objects.DataClasses"#>
<#@ import namespace="System.Linq"#>
<#@ import namespace="System.Reflection"#>
<#@ import namespace="Microsoft.VisualStudio.Web.Mvc.Scaffolding.BuiltIn"#>
<#
MvcTextTemplateHost mvcHost = MvcTemplateHost;
string mvcViewDataTypeGenericString = "<" + mvcHost.ViewDataTypeName + ">";
int CPHCounter = 1;
#>
<#
//The following chained if-statement outputs the user-control needed for a partial view, or
opens the asp:Content tag or html tags used in the case of a master page or regular view page
if(mvcHost.IsPartialView){
#>
```

```
<%@ Control Language="C#"Inherits="System.Web.Mvc.ViewUserControl<#=mvcView-
DataTypeGenericString#>"%>
<#
} else if( mvcHost.IsContentPage) {
#>
<%@ Page Title="" Language="C#"MasterPageFile="<#=mvcHost.MasterPageFile#>"
Inherits="System.Web.Mvc.ViewPage<#=mvcViewDataTypeGenericString#>"%>

<#
    foreach( string cphid in mvcHost.ContentPlaceHolderIDs) {
        if( cphid.Equals("TitleContent",StringComparison.OrdinalIgnoreCase)) {
#>
    <asp:Content ID="Content<#=CPHCounter#>"ContentPlaceHolderID="<#=cphid#>"
runat="server">
        <#=mvcHost.ViewName#>
    </asp:Content>
    <#
            CPHCounter++;
        }
    }
#>
    <asp:Content ID="Content<#=CPHCounter#>"ContentPlaceHolderID="<#=mv-
cHost.PrimaryContentPlaceHolderID#>"runat="server">
    <h2><#=mvcHost.ViewName#></h2>
    <#
} else {
#>
<%@ Page Language="C#"Inherits="System.Web.Mvc.ViewPage<#=mvcViewDataType-
GenericString#>"%>
<!DOCTYPE html>
<html>
<head runat="server">
    <meta name="viewport"content="width=device-width"/>
    <title><#=mvcHost.ViewName#></title>
</head>
<body>
<#
    PushIndent("    ");
}
#>
```

```
<#
if(mvcHost.ReferenceScriptLibraries){
#>
<#
    if(! mvcHost.IsContentPage){
#>
<script src="<%:Url.Content("~/Scripts/jquery-1.8.2.min.js")%>"></script>
<script src="<%:Url.Content("~/Scripts/jquery.validate.min.js")%>"></script>
<script src="<%:Url.Content("~/Scripts/jquery.validate.unobtrusive.min.js")%>"></script>
<#
    }
}
#>
<% using(Html.BeginForm()){%>
    <%:Html.AntiForgeryToken()%>
    <%:Html.ValidationSummary(true)%>
    <fieldset>
        <legend><#=mvcHost.ViewDataType.Name#></legend>
<#
foreach(ModelProperty property in GetModelProperties(mvcHost.ViewDataType)){
    if(! property.IsPrimaryKey && ! property.IsReadOnly && property.Scaffold){
#>
        <div class="editor-label">
<#
        if(property.IsForeignKey){
#>
            <%:Html.LabelFor(model=>model.<#=property.Name#>,"<#=property.AssociationName#>")%>
<#
        }else{
#>
            <%:Html.LabelFor(model=>model.<#=property.Name#>)%>
<#
        }
#>
        </div>
        <div class="editor-field">
<#
        if(property.IsForeignKey){
```

```
#>
            <%:Html.DropDownList("<#=property.Name#>",String.Empty)%>
<#
        } else {
#>
            <%:Html.EditorFor(model=>model.<#=property.Name#>)%>
<#
        }
#>
            <%:Html.ValidationMessageFor(model=>model.<#=property.Name#>)%>
        </div>
<#
    }
}
#>
        <p>
            <input type="submit" value="Create"/>
        </p>
    </fieldset>
<% } %>
<div>
    <%:Html.ActionLink("Back to List","Index")%>
</div>
<#
//The following code closes the asp:Content tag used in the case of a master page and the body and html tags in the case of a regular view page
#>
<#
if(mvcHost.IsContentPage) {
#>
</asp:Content>
<#
    foreach(string cphid in mvcHost.ContentPlaceHolderIDs) {
        if (cphid.Equals("ScriptsSection", StringComparison.OrdinalIgnoreCase) && mvcHost.ReferenceScriptLibraries) {
            CPHCounter++;
#>
<asp:Content ID="Content<#=CPHCounter#>" ContentPlaceHolderID="<#=cphid#>" runat="server">
```

```
            <%:Scripts.Render("~/bundles/jqueryval")%>
        </asp:Content>
        <#
                } else if(!cphid.Equals("TitleContent",StringComparison.OrdinalIgnoreCase)&&!
cphid.Equals(mvcHost.PrimaryContentPlaceHolderID,StringComparison.OrdinalIgnoreCase)){
                    CPHCounter++;
        #>
        <asp:Content ID="Content<#=CPHCounter#>" ContentPlaceHolderID="<#=cphid#>" runat="server">
        </asp:Content>
        <#
                }
            }
        #>
        <#
        } else if(!mvcHost.IsPartialView&&!mvcHost.IsContentPage){
            ClearIndent();
        #>
    </body>
    </html>
    <#
    }
    #>
    <#+
    //Describes the information about a property on the model
    class ModelProperty{
        public string Name{get;set;}
        public string AssociationName{get;set;}
        public string ValueExpression{get;set;}
        public string ModelValueExpression{get;set;}
        public string ItemValueExpression{get;set;}
        public Type UnderlyingType{get;set;}
        public bool IsPrimaryKey{get;set;}
        public bool IsForeignKey{get;set;}
        public bool IsReadOnly{get;set;}
        public bool Scaffold{get;set;}
    }
    //Change this list to include any non-primitive types you think should be eligible for display/edit
    static Type[] bindableNonPrimitiveTypes = new[]{
```

```
            typeof(string),
            typeof(decimal),
            typeof(Guid),
            typeof(DateTime),
            typeof(DateTimeOffset),
            typeof(TimeSpan),
};
//Call this to get the list of properties in the model. Change this to modify or add your
//own default formatting for display values.
List < ModelProperty > GetModelProperties(Type type) {
        List < ModelProperty > results = GetEligibleProperties(type);

        foreach(ModelProperty prop in results) {
                if(prop.UnderlyingType = = typeof(double) || prop.UnderlyingType = = typeof(decimal)) {
                        prop.ModelValueExpression =" String.Format ( \" {0: F} \"," + prop.ModelValueExpression +")";
                }
                else if(prop.UnderlyingType = = typeof(DateTime)) {
                        prop.ModelValueExpression =" String.Format ( \" {0: g} \"," + prop.ModelValueExpression +")";
                }

        }
        return results;
}
//Call this to determine if property has scaffolding enabled
bool Scaffold(PropertyInfo property) {
        foreach(object attribute in property.GetCustomAttributes(true)) {
                var scaffoldColumn = attribute as ScaffoldColumnAttribute;
                if(scaffoldColumn ! = null && ! scaffoldColumn.Scaffold) {
                        return false;
                }
        }
        return true;
}
//Call this to determine if the property represents a primary key. Change the
//code to change the definition of primary key.
bool IsPrimaryKey(PropertyInfo property) {
        if ( string.Equals ( property.Name," id", StringComparison.OrdinalIgnoreCase)) {//EF Code First convention
                return true;
```

```csharp
            }
            if(string.Equals(property.Name, property.DeclaringType.Name + "id", StringComparison.OrdinalIgnoreCase)){//EF Code First convention
                return true;
            }
            foreach(object attribute in property.GetCustomAttributes(true)){
                if(attribute is KeyAttribute){//WCF RIA Services and EF Code First explicit
                    return true;
                }
                var edmScalar = attribute as EdmScalarPropertyAttribute;
                if(edmScalar != null && edmScalar.EntityKeyProperty){//EF traditional
                    return true;
                }
                var column = attribute as ColumnAttribute;
                if(column != null && column.IsPrimaryKey){//LINQ to SQL
                    return true;
                }
            }
            return false;
        }
        //This will return the primary key property name, if and only if there is exactly
        //one primary key. Returns null if there is no PK, or the PK is composite.
        string GetPrimaryKeyName(Type type){
            IEnumerable<string> pkNames = GetPrimaryKeyNames(type);
            return pkNames.Count() == 1? pkNames.First():null;
        }
        //This will return all the primary key names. Will return an empty list if there are none.
        IEnumerable<string> GetPrimaryKeyNames(Type type){
            return GetEligibleProperties(type).Where(mp => mp.IsPrimaryKey).Select(mp => mp.Name);
        }
        //Call this to determine if the property represents a foreign key.
        bool IsForeignKey(PropertyInfo property){
            return MvcTemplateHost.RelatedProperties.ContainsKey(property.Name);
        }
        //A foreign key, e.g. CategoryID, will have a value expression of Category.CategoryID
        string GetValueExpressionSuffix(PropertyInfo property){
            RelatedModel propertyModel;
            MvcTemplateHost.RelatedProperties.TryGetValue(property.Name, out propertyModel);
            return propertyModel != null? propertyModel.PropertyName + "." + propertyModel.DisplayPropertyName:property.Name;
        }
```

```
//A foreign key, e.g. CategoryID, will have an association name of Category
string GetAssociationName(PropertyInfo property) {
    RelatedModel propertyModel;
    MvcTemplateHost.RelatedProperties.TryGetValue(property.Name, out propertyModel);
    return propertyModel != null ? propertyModel.PropertyName : property.Name;
}
//Helper
List<ModelProperty> GetEligibleProperties(Type type) {
    List<ModelProperty> results = new List<ModelProperty>();
    foreach(PropertyInfo prop in type.GetProperties(BindingFlags.Public | BindingFlags.Instance)) {
        Type underlyingType = Nullable.GetUnderlyingType(prop.PropertyType) ?? prop.PropertyType;
        if(prop.GetGetMethod() != null && prop.GetIndexParameters().Length == 0
            && IsBindableType(underlyingType)) {
            string valueExpression = GetValueExpressionSuffix(prop);
            results.Add(new ModelProperty {
                Name = prop.Name,
                AssociationName = GetAssociationName(prop),
                ValueExpression = valueExpression,
                ModelValueExpression = "Model." + valueExpression,
                ItemValueExpression = "item." + valueExpression,
                UnderlyingType = underlyingType,
                IsPrimaryKey = IsPrimaryKey(prop),
                IsForeignKey = IsForeignKey(prop),
                IsReadOnly = prop.GetSetMethod() == null,
                Scaffold = Scaffold(prop)
            });
        }
    }
    return results;
}
//Helper
bool IsBindableType(Type type) {
    return type.IsPrimitive || bindableNonPrimitiveTypes.Contains(type);
}
MvcTextTemplateHost MvcTemplateHost {
    get {
        return (MvcTextTemplateHost)Host;
    }
}
#>
```

6.3.4 记录详细内容显示视图生成模板

记录详细内容显示视图生成模板"Details.tt"的内容如下：

```
<#@ template language = "C#"HostSpecific = "True"#>
<#@ assembly name = "System.ComponentModel.DataAnnotations"#>
<#@ assembly name = "System.Core"#>
<#@ assembly name = "System.Data.Entity"#>
<#@ assembly name = "System.Data.Linq"#>
<#@ import namespace = "System"#>
<#@ import namespace = "System.Collections.Generic"#>
<#@ import namespace = "System.ComponentModel.DataAnnotations"#>
<#@ import namespace = "System.Data.Linq.Mapping"#>
<#@ import namespace = "System.Data.Objects.DataClasses"#>
<#@ import namespace = "System.Linq"#>
<#@ import namespace = "System.Reflection"#>
<#@ import namespace = "Microsoft.VisualStudio.Web.Mvc.Scaffolding.BuiltIn"#>
<#
MvcTextTemplateHost mvcHost = MvcTemplateHost;
string mvcViewDataTypeGenericString = "<" + mvcHost.ViewDataTypeName + ">";
int CPHCounter = 1;
#>
<#
//The following chained if-statement outputs the user-control needed for a partial view, or opens the asp:Content tag or html tags used in the case of a master page or regular view page
if( mvcHost.IsPartialView) {
#>
<%@ Control Language = "C#"Inherits = "System.Web.Mvc.ViewUserControl<# = mvcViewDataTypeGenericString #>"%>
<#
} else if( mvcHost.IsContentPage) {
#>
<%@ Page Title = ""Language = "C#"MasterPageFile = "<# = mvcHost.MasterPageFile #>" Inherits = "System.Web.Mvc.ViewPage<# = mvcViewDataTypeGenericString #>"%>
<#
    foreach( string cphid in mvcHost.ContentPlaceHolderIDs) {
        if( cphid.Equals( "TitleContent",StringComparison.OrdinalIgnoreCase)) {
#>
<asp:Content ID = "Content<# = CPHCounter #>"ContentPlaceHolderID = "<# = cphid #>"runat = "server">
```

```
            <#= mvcHost.ViewName #>
    </asp:Content>
    <#
                CPHCounter++;
            }
        }
    #>
    <asp:Content ID="Content<#= CPHCounter #>" ContentPlaceHolderID="<#= mvcHost.PrimaryContentPlaceHolderID #>" runat="server">
        <h2><#= mvcHost.ViewName #></h2>
    <#
    } else {
    #>
    <%@ Page Language="C#" Inherits="System.Web.Mvc.ViewPage<#= mvcViewDataTypeGenericString #>"%>
    <!DOCTYPE html>
    <html>
    <head runat="server">
        <meta name="viewport" content="width=device-width"/>
        <title><#= mvcHost.ViewName #></title>
    </head>
    <body>
    <#
        PushIndent("    ");
    }
    #>
    <fieldset>
        <legend><#= mvcHost.ViewDataType.Name #></legend>
    <#
    foreach(ModelProperty property in GetModelProperties(mvcHost.ViewDataType)) {
        if(!property.IsPrimaryKey && property.Scaffold) {
    #>
        <div class="display-label">
            <%: Html.DisplayNameFor(model => model.<#= property.ValueExpression #>)%>
        </div>
        <div class="display-field">
            <%: Html.DisplayFor(model => model.<#= property.ValueExpression #>)%>
        </div>
```

```
<#
        }
}
#>
</fieldset>
<p>
<#
string pkName = GetPrimaryKeyName(mvcHost.ViewDataType);
if(pkName != null){
#>
        <%: Html.ActionLink("Edit","Edit",new{id = Model.<#= pkName #>})%> |
        <%: Html.ActionLink("Back to List","Index")%>
<#
} else {
#>
        <%: Html.ActionLink("Edit","Edit",new{/* id = Model.PrimaryKey */})%> |
        <%: Html.ActionLink("Back to List","Index")%>
<#
}
#>
</p>
<#
//The following code closes the asp:Content tag used in the case of a master page and the body and html tags in the case of a regular view page
#>
<#
if(mvcHost.IsContentPage){
#>
</asp:Content>
<#
    foreach(string cphid in mvcHost.ContentPlaceHolderIDs){
        if (!cphid.Equals("TitleContent", StringComparison.OrdinalIgnoreCase) && !cphid.Equals(mvcHost.PrimaryContentPlaceHolderID, StringComparison.OrdinalIgnoreCase)){
            CPHCounter++;
#>
<asp:Content ID="Content<#= CPHCounter #>" ContentPlaceHolderID="<#= cphid #>" runat="server">
</asp:Content>
<#
        }
```

```
        }
#>
<#
} else if( ! mvcHost.IsPartialView && ! mvcHost.IsContentPage) {
    ClearIndent();
#>
</body>
</html>
<#
}
#>
<#+
//Describes the information about a property on the model
class ModelProperty {
    public string Name { get; set; }
    public string AssociationName { get; set; }
    public string ValueExpression { get; set; }
    public string ModelValueExpression { get; set; }
    public string ItemValueExpression { get; set; }
    public Type UnderlyingType { get; set; }
    public bool IsPrimaryKey { get; set; }
    public bool IsForeignKey { get; set; }
    public bool IsReadOnly { get; set; }
    public bool Scaffold { get; set; }
}
//Change this list to include any non-primitive types you think should be eligible for display/edit
static Type[] bindableNonPrimitiveTypes = new[] {
    typeof(string),
    typeof(decimal),
    typeof(Guid),
    typeof(DateTime),
    typeof(DateTimeOffset),
    typeof(TimeSpan),
};
//Call this to get the list of properties in the model. Change this to modify or add your
//own default formatting for display values.
List<ModelProperty> GetModelProperties(Type type) {
    List<ModelProperty> results = GetEligibleProperties(type);
    foreach(ModelProperty prop in results) {
```

```csharp
            if(prop.UnderlyingType == typeof(double) || prop.UnderlyingType == typeof(decimal)){
                prop.ModelValueExpression =" String.Format ( \"{0:F}\"," + prop.ModelValueExpression +")";
            }
            else if(prop.UnderlyingType == typeof(DateTime)){
                prop.ModelValueExpression =" String.Format ( \"{0:g}\"," + prop.ModelValueExpression +")";
            }
        }
    }
    return results;
}
//Call this to determine if property has scaffolding enabled
bool Scaffold(PropertyInfo property){
    foreach(object attribute in property.GetCustomAttributes(true)){
        var scaffoldColumn = attribute as ScaffoldColumnAttribute;
        if(scaffoldColumn != null && !scaffoldColumn.Scaffold){
            return false;
        }
    }
    return true;
}
//Call this to determine if the property represents a primary key. Change the
//code to change the definition of primary key.
bool IsPrimaryKey(PropertyInfo property){
    if ( string.Equals ( property.Name," id", StringComparison.OrdinalIgnoreCase)){//EF Code First convention
        return true;
    }
    if ( string.Equals ( property.Name, property.DeclaringType.Name +" id", StringComparison.OrdinalIgnoreCase)){//EF Code First convention
        return true;
    }
    foreach(object attribute in property.GetCustomAttributes(true)){
        if(attribute is KeyAttribute){//WCF RIA Services and EF Code First explicit
            return true;
        }
        var edmScalar = attribute as EdmScalarPropertyAttribute;
        if(edmScalar != null && edmScalar.EntityKeyProperty){//EF traditional
            return true;
```

```csharp
            }
            var column = attribute as ColumnAttribute;
            if(column != null && column.IsPrimaryKey){//LINQ to SQL
                return true;
            }
        }
        return false;
    }
    //This will return the primary key property name,if and only if there is exactly
    //one primary key. Returns null if there is no PK,or the PK is composite.
    string GetPrimaryKeyName(Type type){
        IEnumerable<string> pkNames = GetPrimaryKeyNames(type);
        return pkNames.Count() == 1? pkNames.First():null;
    }
    //This will return all the primary key names. Will return an empty list if there are none.
    IEnumerable<string> GetPrimaryKeyNames(Type type){
        return GetEligibleProperties(type).Where(mp => mp.IsPrimaryKey).Select(mp => mp.Name);
    }
    //Call this to determine if the property represents a foreign key.
    bool IsForeignKey(PropertyInfo property){
        return MvcTemplateHost.RelatedProperties.ContainsKey(property.Name);
    }
    //A foreign key,e.g. CategoryID,will have a value expression of Category.CategoryID
    string GetValueExpressionSuffix(PropertyInfo property){
        RelatedModel propertyModel;
        MvcTemplateHost.RelatedProperties.TryGetValue(property.Name,out propertyModel);
        return propertyModel != null? propertyModel.PropertyName +"."+ propertyModel.DisplayPropertyName:property.Name;
    }
    //A foreign key,e.g. CategoryID,will have an association name of Category
    string GetAssociationName(PropertyInfo property){
        RelatedModel propertyModel;
        MvcTemplateHost.RelatedProperties.TryGetValue(property.Name,out propertyModel);
        return propertyModel!= null? propertyModel.PropertyName:property.Name;
    }
    //Helper
    List<ModelProperty> GetEligibleProperties(Type type){
        List<ModelProperty> results = new List<ModelProperty>();
        foreach (PropertyInfo prop in type.GetProperties(BindingFlags.Public | BindingFlags.Instance)){
```

```
                Type underlyingType = Nullable.GetUnderlyingType( prop.PropertyType )??
prop.PropertyType;
                if( prop.GetGetMethod( )! = null && prop.GetIndexParameters( ).Length = = 0
&& IsBindableType(underlyingType) ){
                    string valueExpression = GetValueExpressionSuffix( prop );
                    results.Add( new ModelProperty {
                        Name = prop.Name,
                        AssociationName = GetAssociationName( prop ),
                        ValueExpression = valueExpression,
                        ModelValueExpression = "Model. " + valueExpression,
                        ItemValueExpression = "item. " + valueExpression,
                        UnderlyingType = underlyingType,
                        IsPrimaryKey = IsPrimaryKey( prop ),
                        IsForeignKey = IsForeignKey( prop ),
                        IsReadOnly = prop.GetSetMethod( ) = = null,
                        Scaffold = Scaffold( prop )
                    });
                }
            }
        return results;
    }
    //Helper
    bool IsBindableType( Type type ){
        return type.IsPrimitive||bindableNonPrimitiveTypes.Contains(type);
    }
    MvcTextTemplateHost MvcTemplateHost {
        get {
            return( MvcTextTemplateHost )Host;
        }
    }
#>
```

6.3.5 记录编辑视图生成模板

记录编辑视图生成模板"Edit.tt"的内容如下：

```
<#@ template language = "C#"HostSpecific = "True"#>
<#@ assembly name = "System.ComponentModel.DataAnnotations"#>
<#@ assembly name = "System.Core"#>
```

```
<#@ assembly name = "System.Data.Entity"#>
<#@ assembly name = "System.Data.Linq"#>
<#@ import namespace = "System"#>
<#@ import namespace = "System.Collections.Generic"#>
<#@ import namespace = "System.ComponentModel.DataAnnotations"#>
<#@ import namespace = "System.Data.Linq.Mapping"#>
<#@ import namespace = "System.Data.Objects.DataClasses"#>
<#@ import namespace = "System.Linq"#>
<#@ import namespace = "System.Reflection"#>
<#@ import namespace = "Microsoft.VisualStudio.Web.Mvc.Scaffolding.BuiltIn"#>
<#
MvcTextTemplateHost mvcHost = MvcTemplateHost;
string mvcViewDataTypeGenericString = " < " + mvcHost.ViewDataTypeName + " > ";
int CPHCounter = 1;
#>
<#
//The following chained if-statement outputs the user-control needed for a partial view, or opens the asp:Content tag or html tags used in the case of a master page or regular view page
    if(mvcHost.IsPartialView) {
#>
<%@ Control Language = "C#"Inherits = "System.Web.Mvc.ViewUserControl<# = mvcViewDataTypeGenericString #>"%>
<#
} else if(mvcHost.IsContentPage) {
#>
<%@ Page Title = ""Language = "C#"MasterPageFile = "<# = mvcHost.MasterPageFile #>" Inherits = "System.Web.Mvc.ViewPage<# = mvcViewDataTypeGenericString #>"%>
<#
    foreach(string cphid in mvcHost.ContentPlaceHolderIDs) {
        if(cphid.Equals("TitleContent",StringComparison.OrdinalIgnoreCase)) {
#>
<asp:Content ID = "Content<# = CPHCounter #>"ContentPlaceHolderID = "<# = cphid #>"runat = "server" >
<# = mvcHost.ViewName #>
</asp:Content >
<#
            CPHCounter + + ;
        }
    }
#>
```

```
<asp:Content ID="Content<#= CPHCounter #>" ContentPlaceHolderID="<#= mvcHost.PrimaryContentPlaceHolderID #>" runat="server">
    <h2><#= mvcHost.ViewName #></h2>
<#
} else {
#>
<%@ Page Language="C#" Inherits="System.Web.Mvc.ViewPage<#= mvcViewDataTypeGenericString #>" %>
<!DOCTYPE html>
<html>
<head runat="server">
    <meta name="viewport" content="width=device-width"/>
    <title><#= mvcHost.ViewName #></title>
</head>
<body>
<#
    PushIndent("    ");
}
#>
<#
if(mvcHost.ReferenceScriptLibraries) {
#>
<#
    if(!mvcHost.IsContentPage) {
#>
<script src="<%: Url.Content("~/Scripts/jquery-1.8.2.min.js") %>"></script>
<script src="<%: Url.Content("~/Scripts/jquery.validate.min.js") %>"></script>
<script src="<%: Url.Content("~/Scripts/jquery.validate.unobtrusive.min.js") %>"></script>
<#
    }
}
#>
<% using(Html.BeginForm()) {%>
    <%: Html.AntiForgeryToken() %>
    <%: Html.ValidationSummary(true) %>
    <fieldset>
        <legend><#= mvcHost.ViewDataType.Name #></legend>
<#
foreach(ModelProperty property in GetModelProperties(mvcHost.ViewDataType)) {
```

```
            if( property. Scaffold) {
                if( property. IsPrimaryKey) {
#>
                    <% :Html. HiddenFor( model = > model. <# = property. Name #>)%>
<#
                } else if( ! property. IsReadOnly) {
#>
                    < div class = "editor - label">
<#
                        if( property. IsForeignKey) {
#>
                            <% :Html. LabelFor( model = > model. <# = property. Name #>,"<# = property. AssociationName #>")%>
<#
                        } else {
#>
                            <% :Html. LabelFor( model = > model. <# = property. Name #>)%>
<#
                        }
#>
                    </div>
                    < div class = "editor - field">
<#
                        if( property. IsForeignKey) {
#>
                            <% :Html. DropDownList(" <# = property. Name #>",String. Empty)%>
<#
                        } else {
#>
                            <% :Html. EditorFor( model = > model. <# = property. Name #>)%>
<#
                        }
#>
                            <% :Html. ValidationMessageFor( model = > model. <# = property. Name #>)%>
                    </div>
<#
                }
            }
        }
```

```
            #>
                <p>
                    <input type="submit" value="Save"/>
                </p>
            </fieldset>
<%}%>
<div>
    <%: Html.ActionLink("Back to List", "Index")%>
</div>
<#
//The following code closes the asp:Content tag used in the case of a master page and the body and html tags in the case of a regular view page
#>
<#
if(mvcHost.IsContentPage){
#>
</asp:Content>
<#
    foreach(string cphid in mvcHost.ContentPlaceHolderIDs){
        if (cphid.Equals("ScriptsSection", StringComparison.OrdinalIgnoreCase) && mvcHost.ReferenceScriptLibraries){
            CPHCounter++;
#>
<asp:Content ID="Content<#=CPHCounter#>" ContentPlaceHolderID="<#=cphid#>" runat="server">
    <%: Scripts.Render("~/bundles/jqueryval")%>
</asp:Content>
<#
        } else if(!cphid.Equals("TitleContent", StringComparison.OrdinalIgnoreCase) && !cphid.Equals(mvcHost.PrimaryContentPlaceHolderID, StringComparison.OrdinalIgnoreCase)){
            CPHCounter++;
#>
<asp:Content ID="Content<#=CPHCounter#>" ContentPlaceHolderID="<#=cphid#>" runat="server">
</asp:Content>
<#
        }
    }
#>
<#
```

```
} else if( ! mvcHost.IsPartialView && ! mvcHost.IsContentPage) {
    ClearIndent( );
#>
</body>
</html>
<#
}
#>
<#+
//Describes the information about a property on the model
class ModelProperty {
    public string Name{ get; set; }
    public string AssociationName{ get; set; }
    public string ValueExpression{ get; set; }
    public string ModelValueExpression{ get; set; }
    public string ItemValueExpression{ get; set; }
    public Type UnderlyingType{ get; set; }
    public bool IsPrimaryKey{ get; set; }
    public bool IsForeignKey{ get; set; }
    public bool IsReadOnly{ get; set; }
    public bool Scaffold{ get; set; }
}
//Change this list to include any non-primitive types you think should be eligible for display/edit
static Type[ ] bindableNonPrimitiveTypes = new[ ] {
    typeof( string),
    typeof( decimal),
    typeof( Guid),
    typeof( DateTime),
    typeof( DateTimeOffset),
    typeof( TimeSpan),
};
//Call this to get the list of properties in the model. Change this to modify or add your
//own default formatting for display values.
List < ModelProperty > GetModelProperties( Type type) {
    List < ModelProperty > results = GetEligibleProperties( type);
    foreach( ModelProperty prop in results) {
        if( prop.UnderlyingType = = typeof( double) | | prop.UnderlyingType = = typeof( decimal)) {
```

```csharp
                    prop.ModelValueExpression = "String.Format ( \" {0: F} \"," +
prop.ModelValueExpression + ")";
                }
                else if( prop.UnderlyingType == typeof( DateTime ) ){
                    prop.ModelValueExpression = "String.Format ( \" {0: g} \"," +
prop.ModelValueExpression + ")";
                }
            }
        return results;
    }
    //Call this to determine if property has scaffolding enabled
    bool Scaffold( PropertyInfo property ){
        foreach( object attribute in property.GetCustomAttributes( true ) ){
            var scaffoldColumn = attribute as ScaffoldColumnAttribute;
            if( scaffoldColumn != null && ! scaffoldColumn.Scaffold ){
                return false;
            }
        }
        return true;
    }
    //Call this to determine if the property represents a primary key. Change the
    //code to change the definition of primary key.
    bool IsPrimaryKey( PropertyInfo property ){
        if ( string.Equals ( property.Name," id", StringComparison.OrdinalIgnoreCase ) ){//EF
Code First convention
            return true;
        }
        if( string.Equals( property.Name , property.DeclaringType.Name + "id",StringComparison.
OrdinalIgnoreCase ) ){//EF Code First convention
            return true;
        }
        foreach( object attribute in property.GetCustomAttributes( true ) ){
            if( attribute is KeyAttribute ){//WCF RIA Services and EF Code First explicit
                return true;
            }
            var edmScalar = attribute as EdmScalarPropertyAttribute;
            if( edmScalar != null && edmScalar.EntityKeyProperty ){//EF traditional
                return true;
            }
            var column = attribute as ColumnAttribute;
```

```csharp
                if(column != null && column.IsPrimaryKey) {//LINQ to SQL
                    return true;
                }
            }
        return false;
    }
    //This will return the primary key property name, if and only if there is exactly
    //one primary key. Returns null if there is no PK, or the PK is composite.
    string GetPrimaryKeyName(Type type) {
        IEnumerable<string> pkNames = GetPrimaryKeyNames(type);
        return pkNames.Count() == 1 ? pkNames.First() : null;
    }
    //This will return all the primary key names. Will return an empty list if there are none.
    IEnumerable<string> GetPrimaryKeyNames(Type type) {
        return GetEligibleProperties(type).Where(mp => mp.IsPrimaryKey).Select(mp => mp.Name);
    }
    //Call this to determine if the property represents a foreign key.
    bool IsForeignKey(PropertyInfo property) {
        return MvcTemplateHost.RelatedProperties.ContainsKey(property.Name);
    }
    //A foreign key, e.g. CategoryID, will have a value expression of Category.CategoryID
    string GetValueExpressionSuffix(PropertyInfo property) {
        RelatedModel propertyModel;
        MvcTemplateHost.RelatedProperties.TryGetValue(property.Name, out propertyModel);
        return propertyModel != null ? propertyModel.PropertyName + "." + propertyModel.DisplayPropertyName : property.Name;
    }
    //A foreign key, e.g. CategoryID, will have an association name of Category
    string GetAssociationName(PropertyInfo property) {
        RelatedModel propertyModel;
        MvcTemplateHost.RelatedProperties.TryGetValue(property.Name, out propertyModel);
        return propertyModel != null ? propertyModel.PropertyName : property.Name;
    }
    //Helper
    List<ModelProperty> GetEligibleProperties(Type type) {
        List<ModelProperty> results = new List<ModelProperty>();
```

```
                foreach(PropertyInfo prop in type.GetProperties(BindingFlags.Public |
BindingFlags.Instance)){
                Type underlyingType = Nullable.GetUnderlyingType(prop.PropertyType)??
prop.PropertyType;
                if(prop.GetGetMethod()! = null && prop.GetIndexParameters().Length = =0
&& IsBindableType(underlyingType)){
                    string valueExpression = GetValueExpressionSuffix(prop);
                    results.Add(new ModelProperty{
                        Name = prop.Name,
                        AssociationName = GetAssociationName(prop),
                        ValueExpression = valueExpression,
                        ModelValueExpression = "Model." + valueExpression,
                        ItemValueExpression = "item." + valueExpression,
                        UnderlyingType = underlyingType,
                        IsPrimaryKey = IsPrimaryKey(prop),
                        IsForeignKey = IsForeignKey(prop),
                        IsReadOnly = prop.GetSetMethod() = = null,
                        Scaffold = Scaffold(prop)
                    });
                }
            }
        }
        return results;
    }
    //Helper
    bool IsBindableType(Type type){
        return type.IsPrimitive||bindableNonPrimitiveTypes.Contains(type);
    }
    MvcTextTemplateHost MvcTemplateHost{
        get{
            return(MvcTextTemplateHost)Host;
        }
    }
#>
```

6.3.6 记录删除视图生成模板

记录删除视图生成模板"Delete.tt"的内容如下:

```
<#@ template language = "C#"HostSpecific = "True"# >
<#@ assembly name = "System. ComponentModel. DataAnnotations"# >
<#@ assembly name = "System. Core"# >
<#@ assembly name = "System. Data. Entity"# >
<#@ assembly name = "System. Data. Linq"# >
<#@ import namespace = "System"# >
<#@ import namespace = "System. Collections. Generic"# >
<#@ import namespace = "System. ComponentModel. DataAnnotations"# >
<#@ import namespace = "System. Data. Linq. Mapping"# >
<#@ import namespace = "System. Data. Objects. DataClasses"# >
<#@ import namespace = "System. Linq"# >
<#@ import namespace = "System. Reflection"# >
<#@ import namespace = "Microsoft. VisualStudio. Web. Mvc. Scaffolding. BuiltIn"# >
<#
MvcTextTemplateHost mvcHost = MvcTemplateHost;
string mvcViewDataTypeGenericString = " <" + mvcHost. ViewDataTypeName + " >";
int CPHCounter = 1;
#>
<#
//The following chained if – statement outputs the user – control needed for a partial view, or opens the asp:Content tag or html tags used in the case of a master page or regular view page
if( mvcHost. IsPartialView) {
#>
<%@ Control Language = "C#"Inherits = "System. Web. Mvc. ViewUserControl <# = mvcView-DataTypeGenericString #>"%>
<#
} else if( mvcHost. IsContentPage) {
#>
<%@ Page Title = ""Language = "C#"MasterPageFile = " <# = mvcHost. MasterPageFile #>"Inherits = "System. Web. Mvc. ViewPage <# = mvcViewDataTypeGenericString #>"%>
<#
    foreach( string cphid in mvcHost. ContentPlaceHolderIDs) {
        if( cphid. Equals( "TitleContent",StringComparison. OrdinalIgnoreCase)) {
#>
    <asp:Content ID = "Content <# = CPHCounter #>"ContentPlaceHolderID = " <# = cphid #>"runat = "server" >
        <# = mvcHost. ViewName #>
    </asp:Content >
<#
            CPHCounter + + ;
```

```
    }
  }
#>
    <asp:Content ID="Content<#= CPHCounter #>" ContentPlaceHolderID="<#= mvcHost.PrimaryContentPlaceHolderID #>" runat="server">
    <h2><#= mvcHost.ViewName #></h2>
<#
} else {
#>
<%@ Page Language="C#" Inherits="System.Web.Mvc.ViewPage<#= mvcViewDataTypeGenericString #>" %>
<!DOCTYPE html>
<html>
<head runat="server">
    <meta name="viewport" content="width=device-width"/>
    <title><#= mvcHost.ViewName #></title>
</head>
<body>
<#
    PushIndent("    ");
}
#>
<h3>Are you sure you want to delete this?</h3>
<fieldset>
    <legend><#= mvcHost.ViewDataType.Name #></legend>
<#
foreach (ModelProperty property in GetModelProperties(mvcHost.ViewDataType)) {
    if (!property.IsPrimaryKey && property.Scaffold) {
#>
    <div class="display-label">
        <%: Html.DisplayNameFor(model => model.<#= property.ValueExpression #>) %>
    </div>
    <div class="display-field">
        <%: Html.DisplayFor(model => model.<#= property.ValueExpression #>) %>
    </div>
<#
    }
}
```

```
#>
</fieldset>
<% using(Html.BeginForm()){ %>
    <% :Html.AntiForgeryToken()%>
    <p>
        <input type="submit" value="Delete"/> |
        <% :Html.ActionLink("Back to List","Index")%>
    </p>
<% } %>
<#
//The following code closes the asp:Content tag used in the case of a master page and the body and html tags in the case of a regular view page
#>
<#
if(mvcHost.IsContentPage){
#>
</asp:Content>
<#
    foreach(string cphid in mvcHost.ContentPlaceHolderIDs){
        if(! cphid.Equals("TitleContent",StringComparison.OrdinalIgnoreCase) && ! cphid.Equals(mvcHost.PrimaryContentPlaceHolderID,StringComparison.OrdinalIgnoreCase)){
            CPHCounter++;
#>
<asp:Content ID="Content<#= CPHCounter #>" ContentPlaceHolderID="<#= cphid #>" runat="server">
</asp:Content>
<#
        }
    }
#>
<#
} else if(! mvcHost.IsPartialView && ! mvcHost.IsContentPage){
    ClearIndent();
#>
</body>
</html>
<#
}
#>
<#+
```

```csharp
//Describes the information about a property on the model
class ModelProperty {
    public string Name{get;set;}
    public string AssociationName{get;set;}
    public string ValueExpression{get;set;}
    public string ModelValueExpression{get;set;}
    public string ItemValueExpression{get;set;}
    public Type UnderlyingType{get;set;}
    public bool IsPrimaryKey{get;set;}
    public bool IsForeignKey{get;set;}
    public bool IsReadOnly{get;set;}
    public bool Scaffold{get;set;}
}
//Change this list to include any non-primitive types you think should be eligible for display/edit
static Type[] bindableNonPrimitiveTypes = new[]{
    typeof(string),
    typeof(decimal),
    typeof(Guid),
    typeof(DateTime),
    typeof(DateTimeOffset),
    typeof(TimeSpan),
};
//Call this to get the list of properties in the model. Change this to modify or add your
//own default formatting for display values.
List<ModelProperty> GetModelProperties(Type type){
    List<ModelProperty> results = GetEligibleProperties(type);

    foreach(ModelProperty prop in results){
        if(prop.UnderlyingType == typeof(double) || prop.UnderlyingType == typeof(decimal)){
            prop.ModelValueExpression = "String.Format ( \" { 0: F } \"," + prop.ModelValueExpression + ")";
        }
        else if(prop.UnderlyingType == typeof(DateTime)){
            prop.ModelValueExpression = "String.Format ( \" { 0: g } \"," + prop.ModelValueExpression + ")";
        }
    }
    return results;
```

```
}
//Call this to determine if property has scaffolding enabled
bool Scaffold(PropertyInfo property){
    foreach(object attribute in property.GetCustomAttributes(true)){
        var scaffoldColumn = attribute as ScaffoldColumnAttribute;
        if(scaffoldColumn ! = null && ! scaffoldColumn.Scaffold){
            return false;
        }
    }
    return true;
}
//Call this to determine if the property represents a primary key. Change the
//code to change the definition of primary key.
bool IsPrimaryKey(PropertyInfo property){
    if ( string.Equals ( property.Name,"id", StringComparison.OrdinalIgnoreCase)) {//EF Code First convention
        return true;
    }
    if(string.Equals(property.Name,property.DeclaringType.Name + "id",StringComparison.OrdinalIgnoreCase)){//EF Code First convention
        return true;
    }
    foreach(object attribute in property.GetCustomAttributes(true)){
        if(attribute is KeyAttribute){//WCF RIA Services and EF Code First explicit
            return true;
        }
        var edmScalar = attribute as EdmScalarPropertyAttribute;
        if(edmScalar ! = null && edmScalar.EntityKeyProperty){//EF traditional
            return true;
        }
        var column = attribute as ColumnAttribute;
        if(column ! = null && column.IsPrimaryKey){//LINQ to SQL
            return true;
        }
    }
    return false;
}
//This will return the primary key property name,if and only if there is exactly
//one primary key. Returns null if there is no PK,or the PK is composite.
string GetPrimaryKeyName(Type type){
```

```csharp
        IEnumerable<string> pkNames = GetPrimaryKeyNames(type);
        return pkNames.Count() == 1? pkNames.First() :null;
}
//This will return all the primary key names. Will return an empty list if there are none.
IEnumerable<string> GetPrimaryKeyNames(Type type) {
        return GetEligibleProperties(type).Where(mp => mp.IsPrimaryKey).Select(mp => mp.Name);
}
//Call this to determine if the property represents a foreign key.
bool IsForeignKey(PropertyInfo property) {
        return MvcTemplateHost.RelatedProperties.ContainsKey(property.Name);
}
//A foreign key, e.g. CategoryID, will have a value expression of Category.CategoryID
string GetValueExpressionSuffix(PropertyInfo property) {
    RelatedModel propertyModel;
    MvcTemplateHost.RelatedProperties.TryGetValue(property.Name, out propertyModel);
        return propertyModel != null? propertyModel.PropertyName + "." + propertyModel.DisplayPropertyName : property.Name;
}
//A foreign key, e.g. CategoryID, will have an association name of Category
string GetAssociationName(PropertyInfo property) {
    RelatedModel propertyModel;
    MvcTemplateHost.RelatedProperties.TryGetValue(property.Name, out propertyModel);
        return propertyModel != null? propertyModel.PropertyName : property.Name;
}
//Helper
List<ModelProperty> GetEligibleProperties(Type type) {
    List<ModelProperty> results = new List<ModelProperty>();
        foreach(PropertyInfo prop in type.GetProperties(BindingFlags.Public | BindingFlags.Instance)) {
            Type underlyingType = Nullable.GetUnderlyingType(prop.PropertyType) ?? prop.PropertyType;
            if(prop.GetGetMethod() != null && prop.GetIndexParameters().Length == 0
                && IsBindableType(underlyingType)) {
                string valueExpression = GetValueExpressionSuffix(prop);
                results.Add(new ModelProperty {
                    Name = prop.Name,
                    AssociationName = GetAssociationName(prop),
                    ValueExpression = valueExpression,
                    ModelValueExpression = "Model." + valueExpression,
```

```
                    ItemValueExpression = "item." + valueExpression,
                    UnderlyingType = underlyingType,
                    IsPrimaryKey = IsPrimaryKey(prop),
                    IsForeignKey = IsForeignKey(prop),
                    IsReadOnly = prop.GetSetMethod() == null,
                    Scaffold = Scaffold(prop)
                });
            }
        }
        return results;
}
//Helper
bool IsBindableType(Type type){
        return type.IsPrimitive || bindableNonPrimitiveTypes.Contains(type);
}
MvcTextTemplateHost MvcTemplateHost{
        get{
                return(MvcTextTemplateHost)Host;
        }
}
# >
```

另外,系统提供了一个没有模型参数的空视图生成模板"Empty.tt",这里不作介绍。

6.4 自定义 CRUD 模板

系统提供的 CRUD 模板是通用模板,开发人员在此基础上,可以根据实际应用的格式需要修改并定义自己的 CRUD 模板,以便生成所需要的适合开发者要求的运行控制器和相应的视图。在开发"建设工程监理信息系统"项目时,根据用户使用习惯和系统结构组成需求,在原有系统提供的模板基础上设计修改,建立了适合本系统运行要求的 CRUD 模板。

首先,在项目目录下建立子目录"CodeTemplates",然后将 VS 系统安装目录"C:\Program Files\Microsoft Visual Studio 12.0\Common7\IDE\ItemTemplates\CSharp\Web\MVC 4\CodeTemplates"下的两个存储模板的目录"AddController"和"AddView"拷贝到此目录中,如图 6-10 所示。

需要修改并重新定义的模板文件有控制器模板"ControllerWithContext.tt",视图模板文件有"List.tt"、"Create.tt"、"Details.tt"、"Edit.tt"和"Delete.tt"。

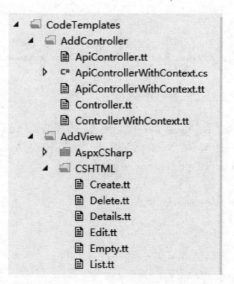

图 6-10 项目中的模板存储目录

6.4.1 自定义控制器模板

常用的控制器模板是"ControllerWithContext.tt",修改后的控制器模板代码内容如下所示:

```
<#@ template language="C#"HostSpecific="True"#>
<#
var Model = (MvcTextTemplateHost)Host;
var routePrefix = String.Empty;
if(!String.IsNullOrWhiteSpace(Model.AreaName)){
    routePrefix = "/" + Model.AreaName;
}
routePrefix += "/" + Model.ControllerRootName + "/";
#>
<#@ import namespace="System.Collections"#>
<#@ import namespace="System.Collections.Generic"#>
<#@ import namespace="System.Data.Objects"#>
<#@ import namespace="System.Linq"#>
using System;
using System.Collections.Generic;
using System.Data;
using System.Data.Entity;
using System.Linq;
```

```
using System.Web;
using System.Web.Mvc;
<# if (MvcTextTemplateHost.NamespaceNeeded(Model.Namespace, Model.ModelType.Namespace)) {#>
using <#= Model.ModelType.Namespace #>;
<#}#>
<# if(Model.ContextType.Namespace != Model.ModelType.Namespace && MvcTextTemplateHost.NamespaceNeeded(Model.Namespace,Model.ContextType.Namespace)){#>
using <#= Model.ContextType.Namespace #>;
<#}#>
namespace <#= Model.Namespace #>
{
<#
    var modelName = Model.ModelType.Name;
    var entitySetName = Model.EntitySetName;
    var modelVariable = modelName.ToLower();
    var entitySetVariable = entitySetName.ToLower();
    var primaryKey = Model.PrimaryKeys[0];
    var lambdaVar = modelVariable[0];
    var isObjectContext = typeof(ObjectContext).IsAssignableFrom(Model.ContextType);
#>
    public class <#= Model.ControllerName #>:Controller
    {
        private <#= Model.ContextType.Name #> db = new <#= Model.ContextType.Name #>();
        private AppService apps = new AppService();
        //验证代码是否存在
        [HttpGet]
        public virtual ActionResult CodeValidate(string <#= primaryKey.Name #> = null)
        {
            bool exists = true;
            var rd = db.<#= entitySetName #>.Find(<#= primaryKey.Name #>);
            if(rd == null) exists = false;
            return Json(!exists,JsonRequestBehavior.AllowGet);
        }
        //
        //GET: <#= routePrefix #>
        public ActionResult Index()
        {
            string pcode = apps.GetSession("projectcode");
```

6.4 自定义 CRUD 模板

```
                    if( string. IsNullOrEmpty( pcode) )
                    {
                        ViewBag. Message ="没有选择工程！请使用[工程选择]功能选择要处
理的工程...";
                        return View("Error");
                    }
                    ViewBag. ModelDisplayName = apps. GetModelDisplayName( typeof( <# = mod-
elName #>));
    <# var includeExpressions ="";
        if( isObjectContext) {
            includeExpressions = String. Join("", Model. RelatedProperties. Values. Select( proper-
ty = > String. Format(". Include( \"{0}\")", property. PropertyName)));
        }
        else {
            includeExpressions = String. Join("", Model. RelatedProperties. Values. Select( proper-
ty = > String. Format(". Include( {0} = > {0}. {1} )",lambdaVar,property. PropertyName)));
        }
    #>
    <# if( ! String. IsNullOrEmpty( includeExpressions) ) {#>
                var <# = entitySetVariable #> = db. <# = entitySetName #>. Where( p = >
p. projectcode = = pcode) <# = includeExpressions #>;
                return View( <# = entitySetVariable #>. ToList( ));
    <# } else {#>
                return View( db. <# = entitySetName #> <# = includeExpressions #>. ToList
( ));
    <# } #>
            }
            //
            //GET: <# = routePrefix #>Details/5
            public ActionResult Details( <# = primaryKey. ShortTypeName #> id = <# = prima-
ryKey. DefaultValue #>)
            {
    <# if( isObjectContext) {#>
                <# = modelName #> <# = modelVariable #> = db. <# = entitySetName #>.
Single( <# = lambdaVar #> = > <# = lambdaVar #>. <# = primaryKey. Name #> = = id);
    <#} else {#>
                <# = modelName #> <# = modelVariable #> = db. <# = entitySetName #>.
Find( id);
    <# } #>
```

```
            if( <# = modelVariable #> = = null)
            {
                return HttpNotFound( );
            }
            return View( <# = modelVariable #>);
        }
        //
        //GET: <# = routePrefix #>Create
        public ActionResult Create( )
        {
<#foreach( var property in Model. RelatedProperties. Values) {#>
            //ViewBag. <# = property. ForeignKeyPropertyName #> = new SelectList( db. <# = property. EntitySetName #>," <# = property. PrimaryKey #>"," <# = property. DisplayPropertyName #>");
    <# } #>
            //return View( );
            // ******************************************
            <# = modelName #>rd = new <# = modelName #>( );
            rd. projectcode = apps. GetSession("projectcode");
            db. <# = entitySetName #>. Add( rd);
            db. SaveChanges( );
            return RedirectToAction("Index");
        }
        //
        //POST: <# = routePrefix #>Create
        [HttpPost]
        public ActionResult Create( <# = modelName #> <# = modelVariable #>)
        {
            if( ModelState. IsValid)
            {
<# if( primaryKey. Type = = typeof( Guid)) {#>
                <# = modelVariable #>. <# = primaryKey. Name #> = Guid. NewGuid( );
    <# } #>
    <# if( isObjectContext) {#>
                db. <# = entitySetName #>. AddObject( <# = modelVariable #>);
    <# } else {#>
                db. <# = entitySetName #>. Add( <# = modelVariable #>);
    <# } #>
```

```
                db.SaveChanges();
                ViewBag.Message = "新增记录存储成功...";
                //return RedirectToAction("Index");
            }
            else
            {
                ViewBag.Message = "数据输入有错误,新增记录存储失败...";
            }
    <# foreach(var property in Model.RelatedProperties.Values){#>
            ViewBag.<#=property.ForeignKeyPropertyName#> = new SelectList(db.<#=property.EntitySetName#>,"<#=property.PrimaryKey#>","<#=property.DisplayPropertyName#>",<#=modelVariable#>.<#=property.ForeignKeyPropertyName#>);
    <#}#>
            return View(<#=modelVariable#>);
        }
        //
        //GET: <#=routePrefix#>Edit/5
        public ActionResult Edit(<#=primaryKey.ShortTypeName#> id = <#=primaryKey.DefaultValue#>)
        {
    <# if(isObjectContext){#>
            <#=modelName#> <#=modelVariable#> = db.<#=entitySetName#>.Single(<#=lambdaVar#> => <#=lambdaVar#>.<#=primaryKey.Name#> == id);
    <#} else {#>
            <#=modelName#> <#=modelVariable#> = db.<#=entitySetName#>.Find(id);
    <#}#>
            if(<#=modelVariable#> == null)
            {
                return HttpNotFound();
            }
    <# foreach(var property in Model.RelatedProperties.Values){#>
            ViewBag.<#=property.ForeignKeyPropertyName#> = new SelectList(db.<#=property.EntitySetName#>,"<#=property.PrimaryKey#>","<#=property.DisplayPropertyName#>",<#=modelVariable#>.<#=property.ForeignKeyPropertyName#>);
    <#}#>
            return View(<#=modelVariable#>);
        }
```

```
            //
            //POST: <# = routePrefix #>Edit/5
            [HttpPost]
            public ActionResult Edit(<# = modelName #> <# = modelVariable #>)
            {
                if(ModelState.IsValid)
                {
<# if(isObjectContext){#>
                    db.<# = entitySetName #>.Attach(<# = modelVariable #>);
                    db.ObjectStateManager.ChangeObjectState(<# = modelVariable #>, EntityState.Modified);
<#}else{#>
                    db.Entry(<# = modelVariable #>).State = EntityState.Modified;
<#}#>
                    db.SaveChanges();
                    ViewBag.Message = "记录编辑存储成功...";
                    //return RedirectToAction("Index");
                }
                else
                {
                    ViewBag.Message = "数据输入有错误,记录编辑存储失败...";
                }
<# foreach(var property in Model.RelatedProperties.Values){#>
                ViewBag.<# = property.ForeignKeyPropertyName #> = new SelectList(db.<# = property.EntitySetName #>,"<# = property.PrimaryKey #>","<# = property.DisplayPropertyName #>",<# = modelVariable #>.<# = property.ForeignKeyPropertyName #>);
<#}#>
                return View(<# = modelVariable #>);
            }
            //
            //GET: <# = routePrefix #>Delete/5
            public ActionResult Delete(<# = primaryKey.ShortTypeName #> id = <# = primaryKey.DefaultValue #>)
            {
<# if(isObjectContext){#>
                <# = modelName #> <# = modelVariable #> = db.<# = entitySetName #>.Single(<# = lambdaVar #> => <# = lambdaVar #>.<# = primaryKey.Name #> == id);
<#}else{#>
```

```
            <# = modelName #> <# = modelVariable #> = db.<# = entitySetName #>.Find(id);
    <# } #>
                if(<# = modelVariable #> == null)
                {
                    return HttpNotFound();
                }
                return View(<# = modelVariable #>);
            }
            //
            //POST: <# = routePrefix #>Delete/5
            [HttpPost,ActionName("Delete")]
            public ActionResult DeleteConfirmed(<# = primaryKey.ShortTypeName #> id)
            {
    <# if(isObjectContext){ #>
                <# = modelName #> <# = modelVariable #> = db.<# = entitySetName #>.Single(<# = lambdaVar #> => <# = lambdaVar #>.<# = primaryKey.Name #> == id);
                db.<# = entitySetName #>.DeleteObject(<# = modelVariable #>);
    <# } else { #>
                <# = modelName #> <# = modelVariable #> = db.<# = entitySetName #>.Find(id);
                db.<# = entitySetName #>.Remove(<# = modelVariable #>);
    <# } #>
                db.SaveChanges();
                return RedirectToAction("Index");
            }
            protected override void Dispose(bool disposing)
            {
                db.Dispose();
                base.Dispose(disposing);
            }
        }
    }
```

其中修改内容有：

(1) 增加方法"CodeValidate"，结合 Ajax，实现对新增记录主键的远程验证功能，对于记录唯一性要求的模型阻止重复记录存在；

(2) 修改英文提示内容为中文内容；

(3) 方法中"return RedirectToAction("Index");"修改为返回到相应的视图。

6.4.2 自定义记录列表显示视图模板

修改后的记录列表显示视图模板的代码内容如下：

```
<#@ template language="C#"HostSpecific="True"#>
<#@ output extension=".cshtml"#>
<#@ assembly name="System.ComponentModel.DataAnnotations"#>
<#@ assembly name="System.Core"#>
<#@ assembly name="System.Data.Entity"#>
<#@ assembly name="System.Data.Linq"#>
<#@ import namespace="System"#>
<#@ import namespace="System.Collections.Generic"#>
<#@ import namespace="System.ComponentModel.DataAnnotations"#>
<#@ import namespace="System.Data.Linq.Mapping"#>
<#@ import namespace="System.Data.Objects.DataClasses"#>
<#@ import namespace="System.Linq"#>
<#@ import namespace="System.Reflection"#>
<#@ import namespace="Microsoft.VisualStudio.Web.Mvc.Scaffolding.BuiltIn"#>
<#
MvcTextTemplateHost mvcHost = MvcTemplateHost;
#>
@model IEnumerable<#=" <" + mvcHost.ViewDataTypeName + ">"#>
<#
//The following chained if-statement outputs the file header code and markup for a partial
view, a content page, or a regular view.
if(mvcHost.IsPartialView){
#>
<#
}else if(mvcHost.IsContentPage){
#>
@{
    ViewBag.Title = ViewBag.ModelDisplayName +"记录数据列表(<#= mvcHost.ViewName#>For Record)";
<#
if(!String.IsNullOrEmpty(mvcHost.MasterPageFile)){
#>
    Layout="<#=mvcHost.MasterPageFile#>";
<#
}
#>
```

```
}
<#
} else {
#>

@ {
    Layout = null;
}
<!DOCTYPE html>

<html>
<head>
    <meta name="viewport" content="width=device-width"/>
    <title><#= mvcHost.ViewName #></title>
</head>
<body>
<#
    PushIndent("    ");
}
#>
<script type="text/javascript">
    $(document).ready(function() {
        var oldback = $("#dtable tr").css("background-color");
        var oldcolor = $("#dtable tr").css("color");
        $("#dtable tr td").click(function() {
            $("#dtable tr").css("background-color", oldback);
            $("#dtable tr").css("color", oldcolor);
            $(this).parent().css("background-color", "#03a5d1");
            $(this).parent().css("color", "white");
        });
    })
</script>
<table id="dtable">
    <caption><h2>@ViewBag.Title</h2></caption>
    <thead>
    <tr>
<#
List<ModelProperty> properties = GetModelProperties(mvcHost.ViewDataType);
foreach(ModelProperty property in properties) {
#>
```

```
            <th>
                @Html.DisplayNameFor(model => model.<#=property.ValueExpression#>)
            </th>
    <#
    }
    #>
            <th class="noprint" style="width:200px;"></th>
        </tr>
    </thead>
    <tbody>
@foreach(var item in Model){
        <tr>
<#
foreach(ModelProperty property in properties){
#>
            <td>
                @Html.DisplayFor(modelItem => <#=property.ItemValueExpression#>)
            </td>
<#
}
string pkName = GetPrimaryKeyName(mvcHost.ViewDataType);
if(pkName != null){
#>
            <td class="noprint">
                @Html.ActionLink("编辑","Edit",new{id=item.<#=pkName#>}) |
                @Html.ActionLink("详细","Details",new{id=item.<#=pkName#>}) |
                @Html.ActionLink("删除","Delete",new{id=item.<#=pkName#>}) |
                @Html.ActionLink("报表","Reports",new{id=item.<#=pkName#>})
            </td>
<#
} else {
#>
            <td class="noprint">
                @Html.ActionLink("编辑","Edit",new{/* id=item.PrimaryKey */}) |
                @Html.ActionLink("详细","Details",new{/* id=item.PrimaryKey */}) |
                @Html.ActionLink("删除","Delete",new{/* id=item.PrimaryKey */}) |
                @Html.ActionLink("报表","Reports",new{/* id=item.PrimaryKey */})
            </td>
<#
}
```

```
                #>
            </tr>
    }
    <tbody>
    </table>
    <p class="noprint">
        @Html.ActionLink("新增","Create",null,new{@class="alinkcss"})
        <button id="print" onclick="window.print();return false;">打印</button>
        @Html.ActionLink("返回","Index","Home",new{Area=""},new{@class="alinkcss"})
    </p>
```

其中修改内容有：

(1) 修改 "ViewBag.Title=" <#=mvcHost.ViewName#>";" 为 "ViewBag.Title = ViewBag.ModelDisplayName +" 记录数据列表（<#= mvcHost.ViewName#>For Record)";";

(2) 修改英文提示内容为中文内容；

(3) 命令和超级链接增加 "class="alinkcss""；

(4) 以表格（Table）方式显示记录；

(5) 删除生成 "ASPX" 视图有关的内容。

6.4.3 自定义新增记录显示视图模板

修改后的新增记录显示视图模板的代码内容如下：

```
<#@ template language="C#" HostSpecific="True"#>
<#@ output extension=".cshtml"#>
<#@ assembly name="System.ComponentModel.DataAnnotations"#>
<#@ assembly name="System.Core"#>
<#@ assembly name="System.Data.Entity"#>
<#@ assembly name="System.Data.Linq"#>
<#@ import namespace="System"#>
<#@ import namespace="System.Collections.Generic"#>
<#@ import namespace="System.ComponentModel.DataAnnotations"#>
<#@ import namespace="System.Data.Linq.Mapping"#>
<#@ import namespace="System.Data.Objects.DataClasses"#>
<#@ import namespace="System.Linq"#>
<#@ import namespace="System.Reflection"#>
<#@ import namespace="Microsoft.VisualStudio.Web.Mvc.Scaffolding.BuiltIn"#>
```

```
<#
MvcTextTemplateHost mvcHost = MvcTemplateHost;
#>
@model <#= mvcHost.ViewDataTypeName #>
<#
//The following chained if-statement outputs the file header code and markup for a partial view, a content page, or a regular view.
if(mvcHost.IsPartialView) {
#>
<#
} else if(mvcHost.IsContentPage) {
#>
@{
    ViewBag.Title = ViewData.ModelMetadata.DisplayName +"增加新记录(<#= mvcHost.ViewName #> New Recoed)";
<#
if(! String.IsNullOrEmpty(mvcHost.MasterPageFile)) {
#>
    Layout = "<#= mvcHost.MasterPageFile#>";
<#
}
#>
}
<#
} else {
#>
@{
    Layout = null;
}
<!DOCTYPE html>
<html>
<head>
    <meta name="viewport" content="width=device-width"/>
    <title><#= mvcHost.ViewName #></title>
</head>
<body>
<#
    PushIndent("    ");
}
#>
```

```
<#
if(mvcHost.ReferenceScriptLibraries){
#>
    <#
        if(! mvcHost.IsContentPage){
#>
<script src="~/Scripts/jquery-1.7.1.min.js"></script>
<script src="~/Scripts/jquery.validate.min.js"></script>
<script src="~/Scripts/jquery.validate.unobtrusive.min.js"></script>
<#
        }
}
#>
@using(Html.BeginForm()){
    @Html.ValidationSummary(true)
    <table id="tt-table">
        <caption><h2>@ViewBag.Title</h2></caption>
<#
foreach(ModelProperty property in GetModelProperties(mvcHost.ViewDataType)){
    if(! property.IsPrimaryKey && ! property.IsReadOnly && property.Scaffold){
#>
        <tr>
            <td class="label-td">
<#
        if(property.IsForeignKey){
#>
                @Html.LabelFor(model => model.<#= property.Name #>,"<#= property.AssociationName #>")
<#
        } else {
#>
                @Html.LabelFor(model => model.<#= property.Name #>)
<#
        }
#>
            </td>
            <td class="field-td">
<#
        if(property.IsForeignKey){
#>
```

```
                        @Html.DropDownList("<#=property.Name#>",String.Empty)
  <#
            } else {
  #>
                        @Html.EditorFor(model => model.<#=property.Name#>)
  <#
            }
  #>
                        @Html.ValidationMessageFor(model => model.<#=property.Name#>)
            </td>
        </tr>
  <#
       }
    }
  #>
        <tr style="border-top:1px solid black;">
            <td class="label-td">操作说明</td>
            <td class="field-td"><#=mvcHost.ViewDataType.Name#>请修改提示内容</td>
        </tr>
    </table>
    <p class="noprint">
        <input type="submit" value="确认存储"/>
        @Html.ActionLink("返回列表","Index",null,new{@class="alinkcss"})
        <span>@ViewBag.Message</span>
    </p>
}
<#
if(mvcHost.IsContentPage && mvcHost.ReferenceScriptLibraries){
#>
@section Scripts{
    @Scripts.Render("~/bundles/jqueryval")
}
<#
}
#>
<script type="text/javascript">
<#
foreach(ModelProperty property in GetModelProperties(mvcHost.ViewDataType))
```

```
         {
              if( property. UnderlyingType = = typeof( DateTime ) )
                   {
#>
              $ ("#<# = property. Name #>"). datepicker( );
<#
                   }
         }
#>
</script>
```

其中修改内容有：

(1) 修改 "ViewBag. Title =" <# = mvcHost. ViewName #>";" 为 "ViewBag. Title = ViewData. ModelMetadata. DisplayName +"增加新记录(<# = mvcHost. ViewName #> New Recoed)";";

(2) 修改英文提示内容为中文内容；

(3) 增加了 "<button id ="print"onclick ="window. print ( ); return false;"> 打印 </button>" 内容；

(4) 以表格（Table）方式显示内容；

(5) 增加日期输入控件与日期输入选择器的绑定代码段；

(6) 删除生成 "ASPX" 视图有关的内容。

6.4.4 自定义记录详细内容显示视图模板

修改后的记录详细内容显示视图模板的代码内容如下：

```
<#@ template language ="C#"HostSpecific ="True"#>
<#@ output extension =". cshtml"#>
<#@ import namespace ="Microsoft. VisualStudio. Web. Mvc. Scaffolding. BuiltIn"#>
<#@ assembly name ="System. ComponentModel. DataAnnotations"#>
<#@ assembly name ="System. Core"#>
<#@ assembly name ="System. Data. Entity"#>
<#@ assembly name ="System. Data. Linq"#>
<#@ import namespace ="System"#>
<#@ import namespace ="System. Collections. Generic"#>
<#@ import namespace ="System. ComponentModel. DataAnnotations"#>
<#@ import namespace ="System. Data. Linq. Mapping"#>
<#@ import namespace ="System. Data. Objects. DataClasses"#>
```

```
<#@ import namespace="System.Linq"#>
<#@ import namespace="System.Reflection"#>
<#@ import namespace="Microsoft.VisualStudio.Web.Mvc.Scaffolding.BuiltIn"#>
<#
MvcTextTemplateHost mvcHost = MvcTemplateHost;
#>
@model <#=mvcHost.ViewDataTypeName#>
<#
//The following chained if-statement outputs the file header code and markup for a partial view, a content page, or a regular view.
if(mvcHost.IsPartialView){
#>
<#
} else if(mvcHost.IsContentPage){
#>
@{
    ViewBag.Title = ViewData.ModelMetadata.DisplayName +"记录数据详细内容(<#=mvcHost.ViewName#>For Record)";
<#
if(!String.IsNullOrEmpty(mvcHost.MasterPageFile)){
#>
    Layout = "<#=mvcHost.MasterPageFile#>";
<#
}
#>
}
<#
} else {
#>
@{
    Layout = null;
}
<!DOCTYPE html>
<html>
<head>
    <meta name="viewport" content="width=device-width"/>
    <title><#=mvcHost.ViewName#></title>
</head>
<body>
<#
```

```
        PushIndent("    ");
    }
#>
<table id="tt-table">
    <caption><h2>@ViewBag.Title</h2></caption>
<#
foreach(ModelProperty property in GetModelProperties(mvcHost.ViewDataType)){
    if(property.IsPrimaryKey && property.Scaffold){
#>
    <tr style="border:none;height:30px;">
        <td class="label-td">
            @Html.DisplayNameFor(model => model.<#= property.ValueExpression #>)
        </td>
        <td class="field-td">
            @Html.DisplayFor(model => model.<#= property.ValueExpression #>)
        </td>
    </tr>
<#
    } else {
#>
    <tr style="border:none;height:30px;">
        <td class="label-td">
            @Html.DisplayNameFor(model => model.<#= property.ValueExpression #>)
        </td>
        <td class="field-td">
            @Html.DisplayFor(model => model.<#= property.ValueExpression #>)
        </td>
    </tr>
<#
    }
}
#>
</table>
<p class="noprint">
    @Html.ActionLink("返回列表","Index",null,new{@class="alinkcss"})
    <button onclick="window.print();return false;">打印输出</button>
</p>
```

其中修改内容有：

（1）修改"ViewBag.Title = " < # = mvcHost.ViewName # >";"为"ViewBag.Title = ViewData.ModelMetadata.DisplayName +"记录数据详细内容（<# = mvcHost.ViewName # > For Record）";";

（2）修改英文提示内容为中文内容；

（3）增加了"< button id = "print" onclick = "window.print（）; return false;" > 打印</button >"内容；

（4）以表格（Table）方式显示内容；

（5）删除了记录编辑链接调用；

（6）删除生成"ASPX"视图有关的内容。

6.4.5 自定义记录编辑显示视图模板

修改后的记录编辑显示视图模板的代码内容如下：

```
<#@ template language = "C#"HostSpecific = "True"# >
<#@ output extension = ".cshtml"# >
<#@ assembly name = "System.ComponentModel.DataAnnotations"# >
<#@ assembly name = "System.Core"# >
<#@ assembly name = "System.Data.Entity"# >
<#@ assembly name = "System.Data.Linq"# >
<#@ import namespace = "System"# >
<#@ import namespace = "System.Collections.Generic"# >
<#@ import namespace = "System.ComponentModel.DataAnnotations"# >
<#@ import namespace = "System.Data.Linq.Mapping"# >
<#@ import namespace = "System.Data.Objects.DataClasses"# >
<#@ import namespace = "System.Linq"# >
<#@ import namespace = "System.Reflection"# >
<#@ import namespace = "Microsoft.VisualStudio.Web.Mvc.Scaffolding.BuiltIn"# >
<#
MvcTextTemplateHost mvcHost = MvcTemplateHost;
# >
@ model  <# = mvcHost.ViewDataTypeName # >
<#
//The following chained if - statement outputs the file header code and markup for a partial view, a content page, or a regular view.
if( mvcHost.IsPartialView) {
# >

<#
```

```
} else if(mvcHost.IsContentPage){
#>
@{
    ViewBag.Title = ViewData.ModelMetadata.DisplayName +"记录数据编辑修改(<#=
mvcHost.ViewName#>Record)";
<#
if(! String.IsNullOrEmpty(mvcHost.MasterPageFile)){
#>
    Layout = "<#=mvcHost.MasterPageFile#>";
<#
}
#>
}
<#
} else {
#>
@{
    Layout = null;
}
<#
}
#>
<!DOCTYPE html>
<html>
<head>
    <meta name="viewport" content="width=device-width"/>
    <title><#=mvcHost.ViewName#></title>
</head>
<body>
<#
    PushIndent("    ");
}
#>
<#
if(mvcHost.ReferenceScriptLibraries){
#>
<#
    if(! mvcHost.IsContentPage){
#>
<script src="~/Scripts/jquery-1.7.1.min.js"></script>
<script src="~/Scripts/jquery.validate.min.js"></script>
<script src="~/Scripts/jquery.validate.unobtrusive.min.js"></script>
```

```
<#
    }
}
#>
@using(Html.BeginForm()){
    @Html.ValidationSummary(true)
    <table id="tt-table">
        <caption><h2>@ViewBag.Title</h2></caption>
<#
foreach(ModelProperty property in GetModelProperties(mvcHost.ViewDataType)){
    if(property.IsPrimaryKey)
    {
#>
        <tr style="border-bottom:1px solid black;">
            <td class="label-td">
                @Html.HiddenFor(model => model.<#=property.Name#>)
                @Html.LabelFor(model => model.<#=property.Name#>)
            </td>
            <td class="field-td">
                @Html.DisplayFor(model => model.<#=property.Name#>)
            </td>
        </tr>
<#
    }
    else
    {
        if(property.IsForeignKey)
        {
#>
        <tr>
            <td class="label-td">
                @Html.LabelFor(model => model.<#=property.Name#>,"<#=property.AssociationName#>")
            </td>
<#
        }
        else
        {
#>
        <tr>
```

```
                    <td class="label-td">
                        @Html.LabelFor(model=>model.<#=property.Name#>)
                    </td>
<#
            }
        if(property.IsForeignKey)
            {
#>
                    <td class="field-td">
                        @Html.DropDownList("<#=property.Name#>",String.Empty)
                    </td>
                </tr>
<#
            }
            else
            {
#>
                    <td class="field-td">
                        @Html.EditorFor(model=>model.<#=property.Name#>)
                        @Html.ValidationMessageFor(model=>model.<#=property.Name#>)
                    </td>
                </tr>
<#
            }
        }
    }
#>
        </table>
        <p class="noprint">
            <input type="submit" value="确认存储"/>
            @Html.ActionLink("返回列表","Index",null,new{@class="alinkcss"})
            @Html.ActionLink("打印输出","Print",null,new{@class="alinkcss"})
            <span>@ViewBag.Message</span>
        </p>
}
<#
if(mvcHost.IsContentPage && mvcHost.ReferenceScriptLibraries){
#>
```

```
@ section Scripts{
    @ Scripts. Render(" ~ /bundles/jqueryval")
}
<#
}
#>
< script type = "text/javascript" >
<#
foreach( ModelProperty property in GetModelProperties( mvcHost. ViewDataType ) )
{
    if( property. UnderlyingType = = typeof( DateTime ) )
    {
#>
        $ ("# <# = property. Name # >"). datepicker( ) ;
<#
    }
}
#>
</script >
```

其中修改内容有：

(1) 修改 " ViewBag. Title =" ViewBag. Title = ViewData. ModelMetadata. DisplayName +"记录数据编辑修改（<# = mvcHost. ViewName # > Record)";";

(2) 修改英文提示内容为中文内容；

(3) 增加了 "@ Html. ActionLink ("打印输出","Print", null, new { @ class ="alinkcss" })";

(4) 以表格（Table）方式显示内容；

(5) 增加日期输入控件与日期输入选择器的绑定代码段；

(6) 删除生成 "ASPX" 视图有关的内容。

6.4.6 自定义记录删除显示视图模板

修改后的记录删除显示视图模板的代码内容如下：

```
< #@ template language = "C#"HostSpecific = "True"# >
< #@ output extension = ". cshtml"# >
< #@ import namespace = "Microsoft. VisualStudio. Web. Mvc. Scaffolding. BuiltIn"# >
< #@ assembly name = "System. ComponentModel. DataAnnotations"# >
```

```
<#@ assembly name = "System.Core"#>
<#@ assembly name = "System.Data.Entity"#>
<#@ assembly name = "System.Data.Linq"#>
<#@ import namespace = "System"#>
<#@ import namespace = "System.Collections.Generic"#>
<#@ import namespace = "System.ComponentModel.DataAnnotations"#>
<#@ import namespace = "System.Data.Linq.Mapping"#>
<#@ import namespace = "System.Data.Objects.DataClasses"#>
<#@ import namespace = "System.Linq"#>
<#@ import namespace = "System.Reflection"#>
<#@ import namespace = "Microsoft.VisualStudio.Web.Mvc.Scaffolding.BuiltIn"#>
<#
MvcTextTemplateHost mvcHost = MvcTemplateHost;
#>
@model <#= mvcHost.ViewDataTypeName #>
<#
//The following chained if - statement outputs the file header code and markup for a partial view, a content page, or a regular view.
if(mvcHost.IsPartialView){
#>


<#
} else if(mvcHost.IsContentPage){
#>
@{
    ViewBag.Title = ViewData.ModelMetadata.DisplayName +"记录删除(<#= mvcHost.ViewName #>)";
<#
if(! String.IsNullOrEmpty(mvcHost.MasterPageFile)){
#>
    Layout = "<#= mvcHost.MasterPageFile#>";
<#
}
#>
}
<#
} else {
#>
@{
    Layout = null;
```

```
    }
    <!DOCTYPE html>
    <html>
    <head>
        <meta name="viewport"content="width=device-width"/>
        <title><#=mvcHost.ViewName#></title>
    </head>
    <body>
    <#
        PushIndent("    ");
    }
    #>
    <table id="tt-table">
        <caption><h2>@ViewBag.Title</h2></caption>
    <#
    foreach(ModelProperty property in GetModelProperties(mvcHost.ViewDataType)){
        if(property.IsPrimaryKey && property.Scaffold)
        {
    #>
            <tr style="border:none;height:30px;">
                <td class="label-td">
                    @Html.DisplayNameFor(model=>model.<#=property.ValueExpression#>)
                </td>
                <td class="field-td">
                    @Html.DisplayFor(model=>model.<#=property.ValueExpression#>)
                </td>
            </tr>
    <#
        }
        else
        {
    #>
            <tr style="border:none;height:30px;">
                <td class="label-td">
                    @Html.DisplayNameFor(model=>model.<#=property.ValueExpression#>)
                </td>
                <td class="field-td">
                    @Html.DisplayFor(model=>model.<#=property.ValueExpression#>)
```

```
            </td>
        </tr>
<#
        }
    }
#>
</table>
<p class = "noprint">
    @using(Html.BeginForm())
    {
        <input type = "submit" value = "确认删除"/>
        @Html.ActionLink("返回列表","Index",null,new{@class = "alinkcss"})
        <span>确认删除当前记录吗?</span>
    }
</p>
```

其中修改内容有:

(1) 修改 " ViewBag. Title = " ViewBag. Title = ViewData. ModelMetadata. DisplayName + "记录删除(<# = mvcHost. ViewName # >)";";

(2) 修改英文提示内容为中文内容;

(3) 以表格(Table)方式显示内容;

(4) 删除生成"ASPX"视图有关的内容。

本章小结

本章首先介绍了 CRUD 模板的概念和作用,然后说明了系统提供的模板内容和开发过程中的常用模板及其运行文件生成的过程,最后说明了在系统模板基础上自定义的 CRUD 模板内容。

7 系统功能设计与实现

"建设工程监理信息系统"项目功能包括工程管理、文档管理、前期准备、施工准备、进度控制、质量控制、造价控制、施工合同其他事项、规程法规、查询统计、系统管理、基础数据等模块,每个模块包括数量不等的子功能。功能的实现是基于实体模型驱动的方式,以 CRUD 模板为模框架,生成原形,然后进行设计实现。鉴于功能设计、开发、实现的方法过程相似,加上篇幅限制,在说明通用功能实现的基础上,选取"工程管理"功能设计与实现进一步具体说明系统功能设计与实现的方法和过程。

本章内容主要有:
7.1 系统主页功能导航
7.2 通用功能导航链接
7.3 工程管理功能实现
7.4 其他功能实现

7.1 系统主页功能导航

系统主页是系统的开始界面,是与用户交互的接口。对于管理信息系统,其主要作用是实现系统功能导航,设计要求简洁明确,操作简单。

7.1.1 主页内容组成结构

系统运行后,其主页显示效果如图 7-1 所示。
主页内容结构由以下部分组成:
(1)公司标识区:显示公司 LOGO、系统名称和系统应用单位名称;
(2)通用功能导航区:显示当前登录用户信息,包括用户标识、用户姓名和用户角色;还有用户切换、工程选择、用户注销、修改密码、系统主页、关于我们这些通用功能链接;
(3)信息提示区:显示当前登录用户所在单位类别及单位名称;当前选择处理的工程编号、名称和注册日期;当前在线用户数量和日期;

7.1 系统主页功能导航

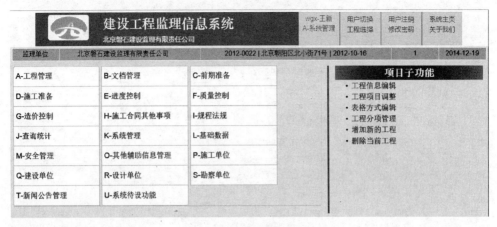

图 7-1 主页显示效果

(4) 主功能导航区：显示系统的全部主功能；
(5) 子功能导航区：显示对应主功能的所有子功能。

7.1.2 主页代码内容

系统主页代码内容文件名称是"_Layout.cshtml"，位于项目根目录（psjlm-vc4 默认）下的"Views/Shared"目录（共享目录）中。此文件并不是直接运行文件，在 ASP.NET MVC 系统中称为布局框架文件，其代码内容如下：

```
@{
    var projectcode = Session["projectcode"];
    var projectname = Session["projectname"];
    var logdate = Session["logdate"];
    var userid = Session["userid"];
    var fullname = Session["fullname"];
    var groupid = Session["groupid"];
    var groupname = Session["groupname"];
    var unittypename = Session["unittypename"];
    var unitname = Session["unitname"];
    var funcode = Session["funcode"];
    var mtitle = Session["mtitle"];
}
<!DOCTYPE html>
<html lang = "utf-8">
<head>
    <meta http-equiv = "Content-Type"content = "text/html;charset = utf-8"/>
    <meta charset = "utf-8"/>
```

```
    <title>@ViewBag.Title - @Bookmvc4.Properties.Resources.ProjectName</title>
    <link href="~/favicon.ico" rel="shortcut icon" type="image/x-icon"/>
    <meta name="viewport" content="width=device-width"/>
    @Scripts.Render("~/bundles/jquery")
    @Scripts.Render("~/bundles/vsdoc")
    @Scripts.Render("~/bundles/modernizr")
    @Scripts.Render("~/bundles/dateZH")
    @Scripts.Render("~/bundles/jqueryui")
    @Styles.Render("~/Content/themes/base/css")
    @Styles.Render("~/Content/css")
    <style type="text/css" media="print">
        .noprint{display:none;}
    </style>
<meta name="description" content="The description of my page"/>
</head>
<body>
    <div class="container" style="background-color:aliceblue;padding:5px;">
        <table class="table hidden-print" id="tb-header">
            <tr>
                <td id="td-logo" style="text-align:center;">
                    <a href="~/Home/Index">
                        <img class="img-rounded" src="~/Images/unitlogo.jpg"/></a>
                </td>
                <td id="td-title">
                    <div class="insetText" style="font-size:2em;">
                        @Bookmvc4.Properties.Resources.ProjectName
                    </div>
                    <div style="font-size:1em;font-family:'Microsoft YaHei';">
                        @Bookmvc4.Properties.Resources.AppliedUnit</div>
                </td>
                <td class="td-link" style="border-right:1px solid blue;">
                    @userid - @fullname <br/>
                    @groupid - @groupname
                </td>
                <td class="td-link" style="border-right:1px solid blue;">
                    @Html.ActionLink("用户切换","Loging","Home",new{Area=""},null) <br/>
                    @Html.ActionLink("工程选择","pSelect","Home",new{Area=""},null)
                </td>
                <td class="td-link" style="border-right:1px solid blue;">
```

```
                            @Html.ActionLink("用户注销","LogOff","Home",new{Area=""},
qexit=true},null) <br/>
                            @Html.ActionLink("修改密码","Updatepwd","Home",new{Area=""},
null)
                        </td>
                        <td class="td-link">
                            @Html.ActionLink("系统主页","Index","Home",new{Area=""},
null) <br/>
                            @Html.ActionLink("关于我们","About","Home",new{Area=""},
null)
                        </td>
                    </tr>
                </table>
                <table id="tb-information" class="table hidden-print">
                    <tr>
                        <td id="td-unittype" style="width:10%;">
                            @unittypename
                        </td>
                        <td id="td-unitname">@unitname</td>
                        <td id="td-project">
                            @projectcode | @projectname | @logdate
                        </td>
                        <td id="td-scount" style="width:10%;">
                            @HttpContext.Current.Application["usercount"]
                        </td>
                        <td id="td-date" style="width:10%;">
                            @DateTime.Today.ToString("yyyy-MM-dd")
                        </td>
                    </tr>
                </table>
                <table id="tb-renderbody" class="table">
                    <tr>
                        <td style="text-align:left;font-size:medium;vertical-align:top;">
                            @RenderBody()
                        </td>
                    </tr>
                </table>
            </div>
            @RenderSection("scripts",required:false)
        </body>
        </html>
```

7.1.3 代码功能说明

系统主页"_Layout.cshtml"是标准的 HTML 格式的文件,在 ASP.NET MVC 框架中,选择后台语言为 C#时,所生成的视图(页面)文件的扩展名称自动定义为".cshtml",而不再分为静态和动态的类别,并使用 Razor 视图引擎语法实现后台功能处理。

"_Layout.cshtml"的功能是实现网站整体布局的统一性,减少 html、head、body 和外部 CSS、JS 引用的大量冗余,其组成代码的功能分为"获取当前登录用户和工程信息"、"引用外部 CSS 和 JS"、"显示标识和通用功能链接"、"显示单位当前工程信息"、"设置@RenderBody()"等部分。

7.1.3.1 获取当前登录用户和工程信息

当前登录用户和工程信息是通过"用户切换"和"工程选择"功能记录在 Session 变量中,在此通过@{……}代码段读取 Session 变量获得相应的用户和当前处理的工程信息,其代码段内容如下:

```
@{
    var projectcode = Session["projectcode"];
    var projectname = Session["projectname"];
    var logdate = Session["logdate"];
    var userid = Session["userid"];
    var fullname = Session["fullname"];
    var groupid = Session["groupid"];
    var groupname = Session["groupname"];
    var unittypename = Session["unittypename"];
    var unitname = Session["unitname"];
    var funcode = Session["funcode"];
    var mtitle = Session["mtitle"];
}
```

在页面需要处使用"@变量名"的语法进行引用。

7.1.3.2 引用外部 CSS 和 JS

外部 CSS 和 JS 引用的语法是"@Styles.Render()"和"@Scripts.Render()",实现了一次引用,重复使用,其代码段内容如下:

```
@Styles.Render("~/Content/themes/base/css")
@Styles.Render("~/Content/css")

@Scripts.Render("~/bundles/jquery")
```

```
@Scripts.Render("~/bundles/vsdoc")
@Scripts.Render("~/bundles/modernizr")
@Scripts.Render("~/bundles/dateZH")
@Scripts.Render("~/bundles/jqueryui")
```

其中，所需要的CSS和JS文件以绑定的方式形成集合，并加以命名，此处则可以通过集合名称加以引用。CSS和JS文件集合的绑定在位于项目子目录"App_Start"文件"BundleConfig.cs"中完成，在项目启动时被执行。"BundleConfig.cs"的内容如下：

```
using System.Web;
using System.Web.Optimization;
namespace Bookmvc4
{
    public class BundleConfig
    {
        //有关Bundling的详细信息，请访问http://go.microsoft.com/fwlink/?LinkId=254725
        public static void RegisterBundles(BundleCollection bundles)
        {
            bundles.Add(new ScriptBundle("~/bundles/jquery").Include(
                "~/Scripts/jquery-{version}.js",
                "~/Scripts/bootstrap.js",
                "~/Scripts/respond.js",
                "~/Scripts/jquery.unobtrusive-ajax.js",
                "~/Scripts/jquery-ui-{version}.js"));
            bundles.Add(new ScriptBundle("~/bundles/vsdoc").Include(
                "~/Scripts/jquery-{version}-vsdoc.js"));
            bundles.Add(new ScriptBundle("~/bundles/jqueryval").Include(
                "~/Scripts/jquery.validate.js",
                "~/Scripts/jquery.validate.unobtrusive.js"));
            bundles.Add(new ScriptBundle("~/bundles/dateZH").Include("~/Scripts/dateZH.js"));
            //使用Modernizr的开发版本进行开发和了解信息。然后，当你做好
            //生产准备时，请使用http://modernizr.com上的生成工具来选择所需的测试。
            bundles.Add(new ScriptBundle("~/bundles/modernizr").Include(
                "~/Scripts/modernizr-*"));
            bundles.Add(new StyleBundle("~/Content/themes/base/css").Include(
                "~/Content/themes/base/*.css"));
            bundles.Add(new StyleBundle("~/Content/css").Include(
```

```
                "~/Content/bootstrap-theme.css",
                "~/Content/bootstrap.css",
                "~/Content/PagedList.css",
                "~/Content/WebSite.css",
                "~/Content/WebGrid.css",
                "~/Content/WordEffects.css"));
        }
    }
}
```

从此处可以了解项目需要的所有外部 CSS 和 JS。

7.1.3.3 显示标识和通用功能链接

系统标识和通用功能链接导航是统一的,并不因页面不同而变化,实现系统标识和通用功能链接导航内容显示的代码如下:

```
<table class="table hidden-print" id="tb-header">
    <tr>
        <td id="td-logo" style="text-align:center;">
            <a href="~/Home/Index">
                <img class="img-rounded" src="~/Images/unitlogo.jpg"/>
            </a>
        </td>
        <td id="td-title">
            <div class="insetText" style="font-size:2em;">
                @Bookmvc4.Properties.Resources.ProjectName
            </div>
            <div style="font-size:1em;font-family:'Microsoft YaHei';">
                @Bookmvc4.Properties.Resources.AppliedUnit
            </div>
        </td>
        <td class="td-link" style="border-right:1px solid blue;">
            @userid - @fullname <br/>
            @groupid - @groupname
        </td>
        <td class="td-link" style="border-right:1px solid blue;">
            @Html.ActionLink("用户切换","Loging","Home",new{Area=""},null)
            <br/>
            @Html.ActionLink("工程选择","pSelect","Home",new{Area=""},null)
        </td>
        <td class="td-link" style="border-right:1px solid blue;">
```

```
@Html.ActionLink("用户注销","LogOff","Home",new{Area=""},qexit=true},null)<br/>
@Html.ActionLink("修改密码","Updatepwd","Home",new{Area=""},null)
        </td>
        <td class="td-link">
@Html.ActionLink("系统主页","Index","Home",new{Area=""},null)
            <br/>
@Html.ActionLink("关于我们","About","Home",new{Area=""},null)
        </td>
    </tr>
</table>
```

7.1.3.4 显示单位和当前工程信息

单位信息是当前登录用户所属单位的类型和单位名称，当前工程是指正在处理的工程，其信息显示实现的代码段内容如下：

```
<table id="tb-information" class="table hidden-print">
    <tr>
        <td id="td-unittype" style="width:10%;">@unittypename</td>
        <td id="td-unitname">@unitname</td>
        <td id="td-project">@projectcode|@projectname|@logdate</td>
        <td id="td-scount" style="width:10%;">
            @HttpContext.Current.Application["usercount"]</td>
        <td id="td-date" style="width:10%;">
            @DateTime.Today.ToString("yyyy-MM-dd")
        </td>
    </tr>
</table>
```

其中包括当前在线用户数量和当前日期的显示。

7.1.3.5 设置@RenderBody()

在 ASP.NET MVC 框架系统中，实现页面内容变化的关键技术是"@RenderBody()"，即子页面内容显示占位符，直接渲染整个 View 到此占位符处并显示。其代码段内容如下：

```
<table id="tb-renderbody" class="table">
    <tr>
        <td style="text-align:left;font-size:medium;vertical-align:top;">
            @RenderBody()
        </td>
    </tr>
</table>
```

占位符内容少但作用大。在 Razor 视图中，除"@ RenderBody( )"占位符外，还有以下格式的占位符可用：@ RenderPage( )方法——渲染指定的页面到占位符处；@ RenderSection 方法——声明一个占位符；@ section 标记——对@ RenderSection 方法声明的占位符进行实现。

7.1.4 @RenderBody( )方法的实现

"_Layout. cshtml"位于项目目录"~/Views/Shared/"中，是建立项目时系统自动默认建立的页面布局文件，在此项目不做更改，直接使用（当然如果需要可以更改为其他名称），其内容如 7.1.2 和 7.1.3 所述。

在项目目录"~/Views/"中有一个文件，名称为"_ViewStart. cshtml"，同样由系统建立项目时自动建立，其内容如下：

```
@{
    Layout = "~/Views/Shared/_Layout. cshtml";
}
```

其为告知项目，所使用布局页面是"~/Views/Shared/_Layout. cshtml"，这样在执行某个页面时就会将此页面内容渲染到"@ RenderBody( )"所在的位置处。

现以请求"Home/Index"页面为例，说明"@ RenderBody( )"方法的实现。首先发出请求"http：//localhost/Home/Index"；查找控制器"Home"；在控制器"Home"中查找方法"Index"；在"Index"方法中有"return View( )"命令，返回"Index. cshtml"的内容，并渲染到"@ RenderBody( )"所出现的位置处。

因为"@ RenderBody( )"在"_Layout. cshtml"中是唯一的，所以不会出现歧义的情况。当然在建立视图对话框中，需要勾选"使用布局页（U）"选项，如图 7-2 所示。

这里默认的布局页就是由"_ViewStart. cshtml"所指定的"_Layout. cshtml"，生成的视图（View）代码内容格式如下所示：

```
@{
    ViewBag. Title = "页面标题";
}
<!—页面内容—>
```

"页面内容"就是根据实际需要所编写的内容。

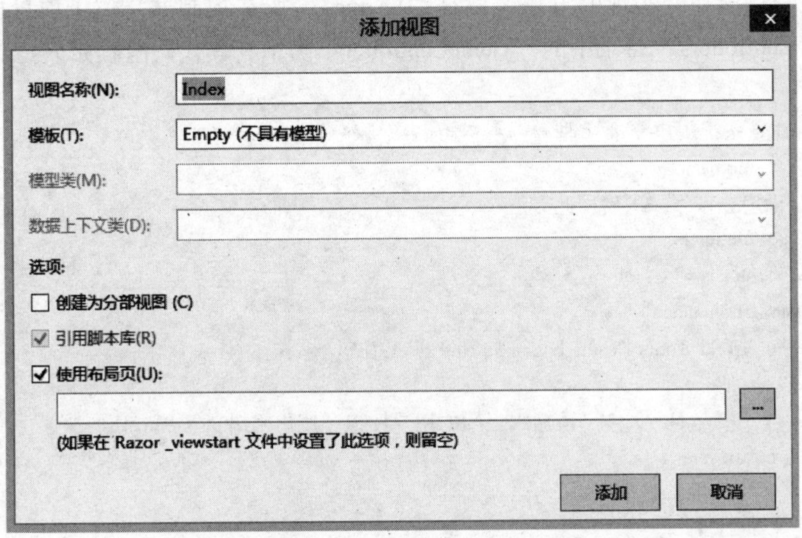

图 7-2 添加视图对话框

7.2 通用功能导航链接

主页上的通用功能导航链接包括"用户切换"、"工程选择"、"用户注销"、"修改密码"、"系统主页"和"关于我们",这些功能对于登录和非登录用户都是可见的。

7.2.1 用户切换

用户切换实现系统已注册用户的登录或重新登录功能,随时可以被登录和未登录用户调用。用户通过登录,确定用户的角色,并完成用户标识、用户名称、单位类型、单位名称等和用户相关的信息记录。用户切换的运行界面如图 7-3 所示。

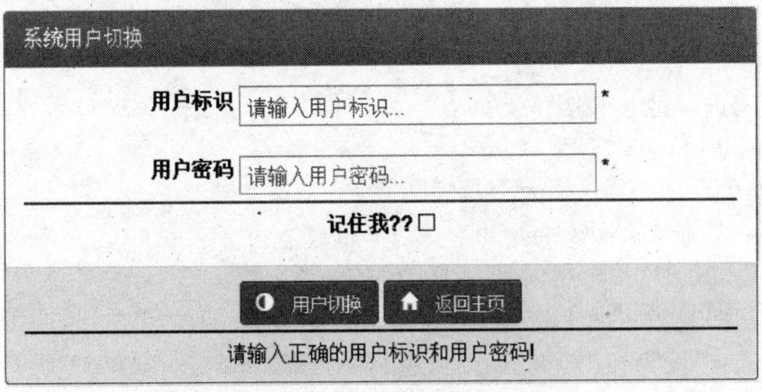

图 7-3 用户切换运行界面

用户切换功能实现的方法名称是"Loging",所在控制器是位于项目根目录下的"Controllers"目录中的"HomeController"类中,其代码内容如下:

```
#region 系统用户登录管理 ******************
/// < summary >
///用户登录第一次启用此方法
/// </ summary >
/// < returns > </ returns >
[AllowAnonymous]
public virtual ActionResult Loging( string returnUrl)
{
    ViewBag. Message ="请输入正确的用户标识和用户密码!" + returnUrl;
    return View();
}
/// < summary >
///用户输入登录信息并提交后所启用的方法
/// </ summary >
/// < param name ="fc" > </ param >
/// < returns > </ returns >
[HttpPost]
[AllowAnonymous]
public virtual ActionResult Loging( KUserLoging ulm, string returnUrl)
{
    string uid = Convert. ToString( ulm. userid);
    string pwd = Convert. ToString( ulm. password);
    var rd = db. LUserStaffs. SingleOrDefault( u = > u. userid = = uid &&
                                                u. password = = pwd);
    if( rd = = null)
    {
        ViewBag. Message ="登录失败!用户名称或口令输入有误!";
        return View( ulm);
    }
    //代码进行到此,说明登录成功
    rd. logincount + = 1;
    db. SaveChanges();//更新登录次数
    #region 保存变量,供其他应用(页面)使用
    apps. SetSession("userid", uid);
    Session. Add("fullname", rd. fullname);
    apps. SetAppState("usercount", Convert. ToString( OnlineCount( uid, true)));
    //用户角色信息
```

```
if( String. IsNullOrEmpty( rd. groupid) )
{
    Session. Add("groupid","");
    Session. Add("groupname","");
}
else
{
    Session. Add("groupid",rd. groupid??"");
    Session. Add("groupname",rd. KGroupList. groupname??"");
}
//用户单位和单位类型信息
if( String. IsNullOrEmpty( rd. unitcode) )
{
    Session. Add("unitcode","");
    Session. Add("unitname","");
    Session. Add("unittypename","");
}
else
{
    Session. Add("unitcode",rd. unitcode??"");
    Session. Add("unitname",rd. LUnitList. unitname??"");
Session. Add("unittypename",rd. LUnitList. LUnitType. unittypename??"");
}
#endregion
//修改系统登录信息
FormsAuthentication. SetAuthCookie( uid,true,FormsAuthentication.
                                          FormsCookiePath);
#region 写入日志
    KLoginList ul = new KLoginList( );
    ul. userid = uid;
    ul. event1 = apps. GetSession("groupname");
    ul. logintime = DateTime. Now;
    ul. actiontype = "in";
    ul. hostname = Dns. GetHostName( );
    db. KLoginLists. Add( ul);
    db. SaveChanges( );
#endregion
if( ! String. IsNullOrEmpty( returnUrl) )
{
    return Redirect( returnUrl);
```

```
        }
        else
        {
            return RedirectToAction("Index","Home");
        }
    }
#endregion 系统用户登录管理结束
```

这里有两个"Loging"方法：第一个是用户第一次调用用户切换功能时，显示用户登录界面的方法，没有实体模型参数；第二个是用户完成用户标识和用户密码输入并提交后调用的方法，传递的参数类型实例变量名称是 KUserLoging ulm，即用户登录实体模型。"Loging"方法完成以下功能：

（1）进入用户登录界面；
（2）提交并接受用户信息，传递给实例变量 ulm；
（3）根据用户标识和用户密码检索相应的记录，如果不存在，返回登录界面，否则，进行下面的工作；
（4）修改用户登录次数，并存入数据库；
（5）将用户信息存入相应的 Session 变量中；
（6）修改系统登录上下文中的用户信息；
（7）将登录信息写入日志；
（8）检索用户单位信息并存入相应的 Session 变量中；
（9）返回系统主页。

方法对应的视图文件名称是"Loging.cshtml"，位于项目根目录下的"Views/Home"目录中，其代码内容如下：

```
@model psjlmvc4.Models.KUserLoging
@{
    ViewBag.Title ="系统用户切换";
}
<div class ="center-block">
    @using(Html.BeginForm())
    {
        <div class ="panel panel-primary"
            style ="width:600px;text-align:center;margin:auto;">
            <div class ="panel-heading"style ="text-align:left;">
                @ViewBag.Title
            </div>
```

```
            <div class="panel-body">
                <div>
                    @Html.LabelFor(m => m.userid)
                    @Html.TextBoxFor(m => m.userid,
new{placeholder="请输入用户标识...",style="opacity:0.5;"})
                    <span class="message">*</span>
                </div>
                <br/>
                <div>
                    @Html.LabelFor(m => m.password)
                    @Html.PasswordFor(m => m.password,
new{placeholder="请输入用户密码...",style="opacity:0.5;"})
                    <span class="message">*</span>
                </div>
                <hr/>
                <div>
                    @Html.LabelFor(m => m.rememberme)
                    @Html.CheckBoxFor(m => m.rememberme)
                </div>
            </div>
            <div class="panel-footer">
                <button type="submit"
class="btn btn-primary glyphicon glyphicon-adjust">
                    用户切换
                </button>
                @Html.ActionLink("返回主页","Index",null,
new{@class="btn btn-primary glyphicon glyphicon-home"})
                <hr/>
                <div class="message">@ViewBag.Message</div>
            </div>
        </div>
    }
</div>
@section Scripts{
    @Scripts.Render("~/bundles/jqueryval")
}
```

7.2.2 工程选择

工程选择完成要管理工程的选择任务,是工程监理工作管理操作对象,所有监理任务必须以工程为基础开展，因此工程选择是系统进行的前提。工程选择功能只有登录用户有权使用，其运行界面如图7-4所示。

图7-4 工程选择界面

工程选择界面分为两个部分：工程信息列表和工程选择条件设置。

实现工程选择功能的方法名称是"pSelect"，所属控制器是"HomeController"类，位于项目根目录下的"Controllers"目录中，其代码内容如下：

```
#region 工程选择处理方法
/// < summary >
///Submit:fc. AllKeys. Length > 0
/// </ summary >
/// < param name = "fc" > </param >
/// < returns > </returns >
//[ MyAuth( Roles = "ABCDEF") ]
public virtual ActionResult pSelect( FormCollection fc)
{
    string uid = apps. GetSession("userid") ;
    if( String. IsNullOrEmpty( HttpContext. User. Identity. Name) )
    {
        ViewBag. Message ="没有用户登录！请先登录系统…";
        return View("Error") ;
    }
```

```csharp
string pcode = "";
string pname = "";
DateTime pdate1 = DateTime.Today;
DateTime pdate2 = DateTime.Today;
int prows = 20;
//====================================================
if(fc.AllKeys.Count() > 0)
{
    apps.SetSession("pcode",fc["pcode"]);
    apps.SetSession("pname",fc["pname"]);
    apps.SetSession("pdate1",Convert.ToString(fc["pdate1"]));
    apps.SetSession("pdate2",Convert.ToString(fc["pdate2"]));
    apps.SetSession("prows",Convert.ToString(fc["prows"]));
}
pcode = apps.GetSession("pcode");
pname = apps.GetSession("pname");
pdate1 = String.IsNullOrEmpty(apps.GetSession("pdate1"))?
        pdate1:Convert.ToDateTime(apps.GetSession("pdate1"));
pdate2 = String.IsNullOrEmpty(apps.GetSession("pdate2"))?
        pdate1:Convert.ToDateTime(apps.GetSession("pdate2"));
prows = String.IsNullOrEmpty(apps.GetSession("prows"))?
        prows:Convert.ToInt32(apps.GetSession("prows"));
ViewBag.pcode = pcode;
ViewBag.pname = pname;
ViewBag.pdate1 = pdate1.ToString("yyyy-MM-dd");
ViewBag.pdate2 = pdate2.ToString("yyyy-MM-dd");
ViewBag.prows = prows;
ViewBag.Message = ViewBag.pdate1 + "===" + ViewBag.pdate2;
if("ACE".IndexOf(apps.GetSession("groupid")) >= 0)
{
    //用户角色是"A","E","C"时检索工程
    var rd = from p in db.AProjectLists
            where p.projectcode.Contains(pcode)&&
            p.projectname.Contains(pname)
            select p;
    rd = rd.Where(p => p.logdate >= pdate1 && p.logdate <= pdate2);
    return View(rd.ToList());
}
else
{
```

```
            //根据 USERID 检索工程
            var rd = from p in db. AProjectLists
            join s in db. CAttendStaffs on p. projectcode equals s. projectcode
                where s. userid = = uid
                select p;
            rd = from p in rd
                where p. projectcode. Contains( pcode) &&
                    p. projectname. Contains( pname)
                select p;
            rd = rd. Where( p = > p. logdate > = pdate1 && p. logdate < = pdate2);
            return View( rd. ToList( ) );
        }
    }
```

"pSelect"方法完成以下任务：

（1）根据"HttpContext. User. Identity. Name"，判断是否为注册用户，不是则显示系统运行状态提示页"Error"，提示此功能需要登录后方可使用；

（2）定义检索变量；

（3）通过"FormCollection fc"集合接受用户输入的检索条件数据；

（4）根据集合变量设置相应的检索变量值和 ViewBag 动态变量；

（5）根据用户的角色，按检索变量完成工程检索；

（6）检索结果返回工程检索界面，并显示。

在工程检索界面上，用户通过每条记录的"选择"链接完成工程的选择。"选择"链接调用方法是"pSelected"，完成相应 Session 变量的设置（记录工程的编号、名称和注册日期）。"pSelected"是控制器"HomeController"类中的方法，其代码内容如下：

```
/// < summary >
///工程选择后保存工程信息
/// </ summary >
/// < param name ="pcode" > </ param >
/// < param name ="pname" > </ param >
/// < param name ="pdate" > </ param >
/// < returns > </ returns >
public virtual ActionResult pSelected( string pcode, string pname, string pdate)
{
    ViewBag. Message ="所选择工程信息:" + pcode + "||" + pname + "||" + pdate;
```

```csharp
apps.SetSession("projectcode",pcode);
apps.SetSession("projectname",pname);
apps.SetSession("logdate",pdate);
//取参与人员名称
var rd1 = db.CAttendStaffs.Where(c = > c.projectcode = = pcode);
foreach(var item in rd1.ToList())
{
    try
    {
        switch(item.LUserStaff.groupid)
        {
            case "D":
                apps.SetSession("supervisor",
                    item.LUserStaff.fullname);
                break;//监理工程师
            case "F":
                apps.SetSession("generalsupervisor",
                    item.LUserStaff.fullname);
                break;//总监理工程师
            case "H":
                apps.SetSession("superman",
                    item.LUserStaff.fullname);
                break;//监理员
            case "O":
                apps.SetSession("techman",
                    item.LUserStaff.fullname);
                break;//技术负责人
            case "P":
                apps.SetSession("projectmanager",
                    item.LUserStaff.fullname);
                break;//项目经理
        }
    }
    catch{}
}
//取参与单位名称
var rd3 = db.CAttendUnits.Where(c = > c.projectcode = = pcode);
foreach(var item in rd3.ToList())
{
    try
```

```
            {
                        switch(item.LUnitList.unittypecode)
                        {
                            case "A":
                                apps.SetSession("supervisionunit",
                                    item.LUnitList.unitname);
                                break;//监理单位名称
                            case "B":
                                apps.SetSession("constructunit",
                                    item.LUnitList.unitname);
                                break;//承包单位名称
                            case "C":
                                apps.SetSession("constructunit2",
                                    item.LUnitList.unitname);
                                break;//分包单位名称
                            case "D":
                                apps.SetSession("buildunit",
                                    item.LUnitList.unitname);
                                break;//建设单位名称
                            case "E":
                                apps.SetSession("designunit",
                                    item.LUnitList.unitname);
                                break;//设计单位名称
                            case "F":
                                apps.SetSession("surveyunit",
                                    item.LUnitList.unitname);
                                break;//勘察单位名称
                        }
            }
            catch{}
            return RedirectToAction("Index","Home");
}
```

"pSelected"方法完成的其他任务还有设置工程相关的管理人员和参与单位信息记录变量。因为不需要显示，所以"pSelected"方法没有对应的视图。

7.2.3 用户注销

用户注销功能的任务是使当前用户退出系统、清除与用户有关的Session变量信息,并写入日志记录。用户注销的方法名称是"LogOff",所属控制器是"HomeController"类,位于项目根目录下的"Controllers"目录,没有对应的视图。"LogOff"方法的代码内容如下:

```
public virtual ActionResult LogOff(bool qexit = false)
{
    var uid = apps.GetSession("userid");
    if(String.IsNullOrEmpty(uid))
    {
        ViewBag.Message = "目前没有用户登录!";
        return View("Error");
    }
    //写入日志
    var ul = db.KLoginLists.Where(p = > p.userid = = uid &&
        p.actiontype = = "in").ToList().LastOrDefault();
    if(ul! = null)
    {
        ul.logofftime = DateTime.Now;
        ul.actiontype = "off";
        db.SaveChanges();
    }
    FormsAuthentication.SignOut();
    apps.SetAppState("usercount",
      Convert.ToString(OnlineCount(uid,false)));
    Session.Abandon();
    //强制取消当前会话,执行 SESSION_END 事件,关键所在!
    if(qexit) return RedirectToAction("Index","Home");
    else return Content("安全退出");
}
```

7.2.4 修改密码

修改密码的功能是完成当前登录用户自行修改自己的登录密码的任务,因此,只有登录用户可用此功能。修改密码功能的运行界面如图 7-5 所示。

修改密码的方法名称是"Updatepw",所属控制器是"HomeController"类,位于项目根目录下的"Controllers"目录中。"Updatepw"方法的代码内容如下:

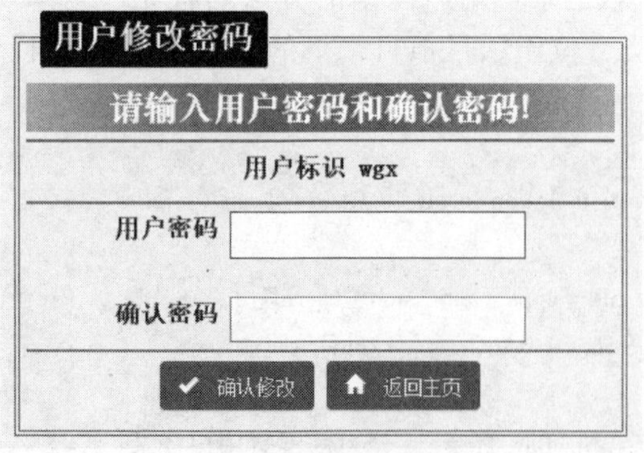

图 7-5 修改密码界面

```
[Authorize]
public virtual ActionResult Updatepwd(KUserAdd up)
{
    string uid = apps.GetSession("userid");
    if(String.IsNullOrEmpty(uid))
    {
        ViewBag.Message = "当前没有用户登录!";
        return View("Error");
    }
    up.userid = uid;
    if(String.IsNullOrEmpty(up.password) ||
        String.IsNullOrEmpty(up.confirmpassword))
    {
        ViewBag.Message = "请输入用户密码和确认密码!";
        return View(up);
    }
    if(up.password == up.confirmpassword)
    {
        var rd = db.LUserStaffs.Find(uid);
        rd.password = up.password;
        db.SaveChanges();
        ViewBag.Message = "密码修改完毕!";
    }
    else
    {
        ViewBag.Message = "用户密码和确认密码不一致!";
    }
    return View(up);
}
```

"Updatepw"方法完成的任务有：
（1）判断是否为已登录用户；
（2）接受用户输入；
（3）判断用户输入是否有空值；
（4）判断用户密码和确认密码是否一致；
（5）修改密码并存入数据库。

修改密码方法对应的视图文件是"Updatepw.cshtml"，存储于项目根目录下的"Views/Home"目录中，其代码内容如下：

```
@model psjlmvc4.Models.KUserAdd
@{
    ViewBag.Title = "用户修改密码";
}
<div class = "row">
    <div class = "col-lg-offset-3">
        @using(Html.BeginForm("Updatepwd","Home",FormMethod.Post))
        {
            <fieldset style = "width:50%;text-align:center;">
                <legend>@ViewBag.Title</legend>
                <h2>@ViewBag.Message</h2>
                <hr/>
                @Html.LabelFor(model => model.userid)
                @Html.DisplayFor(model => model.userid)
                @Html.HiddenFor(model => model.userid)
                <hr/>
                @Html.LabelFor(model => model.password)
                @Html.EditorFor(model => model.password)
                <br/><br/>
                @Html.LabelFor(model => model.confirmpassword)
                @Html.EditorFor(model => model.confirmpassword)
                <hr/>
                <p>
                    <button type = "submit"class = "btn btn-primary glyphicon glyphicon-ok">确认修改</button>
                    @Html.ActionLink("返回主页","Index","Home",null,new{@class = "btn btn-primary glyphicon glyphicon-home"})
                </p>
            </fieldset>
        }
    </div>
</div>
```

参数的传递是通过实体模型"KUserAdd"实现的。

7.2.5 系统主页

系统主页的任务是显示系统功能导航,为用户完成系统所提供功能的接口,其运行界面如图 7-6 所示。

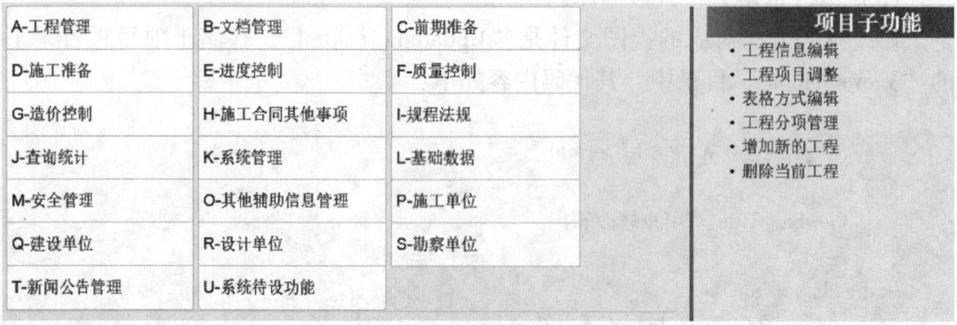

A-工程管理	B-文档管理	C-前期准备	项目子功能
D-施工准备	E-进度控制	F-质量控制	· 工程信息编辑 · 工程项目调整 · 表格方式编辑 · 工程分项管理 · 增加新的工程 · 删除当前工程
G-造价控制	H-施工合同其他事项	I-规程法规	
J-查询统计	K-系统管理	L-基础数据	
M-安全管理	O-其他辅助信息管理	P-施工单位	
Q-建设单位	R-设计单位	S-勘察单位	
T-新闻公告管理	U-系统待设功能		

图 7-6 系统主页显示界面

界面显示内容分为两个部分:左边是系统主功能列表;右边是对应主功能项目的子功能列表。系统主页的方法名称是"Index",所属控制器是"HomeController"类,位于项目根目录下的"Controllers"目录中。"Index"方法的代码内容如下:

```
public virtual ActionResult Index()
{
    ViewBag.Message ="系统门户首页";
    var rd = db.KFunLists.Where(k => k.funcode.Trim().Length == 1);
    return View(rd.ToList());
}
```

系统主页方法的任务是检索系统主功能项目并传递到对应的视图进行显示。系统主页方法对应的视图文件名称是"Index.cshtml",存储于项目根目录下的"Views/Home"目录中,其代码内容如下:

```
@model IEnumerable<psjlmvc4.Models.KFunList>
@{
    ViewBag.Title ="系统主页";
    string fc = Session["funcode"] == null?
        "A":Session["funcode"].ToString();
}
```

```html
<div class="row" style="font-size:large;">
    <div class="col-sm-8" style="border-right:4px double red;">
        <ul class="list-group list-inline" id="ul-mainmenu">
            @foreach(var item in Model)
            {
                <li class="list-group-item">
                    @Ajax.ActionLink(item.funname,"SubMenuListPartial","Home", new{fcode=item.funcode},
                        new AjaxOptions{UpdateTargetId="div-submenulist"})
                </li>
            }
        </ul>
    </div>
    <div class="col-sm-4">
        <h2>项目子功能</h2>
        <div id="div-submenulist" style="overflow-y:auto;height:300px;text-align:left;">
            @{Html.RenderAction("SubMenuListPartial","Home",new{fcode=fc});}
        </div>
    </div>
</div>
<!--*******************************************-->
<style type="text/css">
    #ul-mainmenu li a{
        width:200px;
        display:inline-block;
    }
</style>
<script type="text/javascript">
    $(document).ready(function(){
        var licolor = $("#ul-mainmenu li a").css("background-color");
        $("#ul-mainmenu li a").click(function(){
            $("#ul-mainmenu li a").css("background-color",licolor);
            $(this).css("background-color","red");
        })
    })
</script>
```

对应主功能项目的子功能的显示是通过 Ajax 方式实现，其实现的代码段如下所示：

```
@Ajax.ActionLink(item.funname,"SubMenuListPartial","Home",
    new{fcode = item.funcode},
    new AjaxOptions{UpdateTargetId = "div-submenulist"})
```

此代码的功能如下：
（1）连接并调用控制器"Home"中子功能项目目录显示的方法"SubMenuListPartial"；
（2）传送子功能代码"Funcode"给方法变量"fcode"；
（3）将子功能项目目录显示到"id = div - submenulist"指定的 div 区域。
子功能项目目录显示方法"SubMenuListPartial"的代码内容如下：

```
//显示子菜单功能（分部视图）
//[HttpPost]
public ActionResult SubMenuListPartial(string fcode = "")
{
    apps.SetSession("funcode",fcode);
    var gid = apps.GetSession("groupid");
    var rd = from g in db.KGroupFuns
        where g.groupid == gid && g.funcode.Substring(0,1) == fcode &&
        g.funcode.Length > 1
        from f in db.KFunLists
        where f.funcode == g.funcode
        orderby f.funcode
        select f;
    return PartialView(rd.ToList());
}
```

方法的关键是根据主功能编号检索相应的功能记录，并以分部视图的方式返回检索结果到显示视图中。子功能项目目录显示方法"SubMenuListPartial"对应的视图文件名称是"SubMenuListPartial.cdhtml"，是一个分部（局部）类型的视图，同样存储于项目根目录下的"Views/Home"目录中，其代码内容如下：

```
@model IEnumerable<psjlmvc4.Models.KFunList>
<div style = "text-align:center;font-family:'Microsoft YaHei';
    font-size:20px;background-color:#7db9e8;">
```

```
        @ViewBag.mtitle
</div>
<ul id="submenu">
    @foreach(var item in Model)
    {
        <li>
            @Html.ActionLink(item.funname, item.actionname,
                item.controllername,
                new{Area=@item.funcode.Substring(0,1)+"Area"},
                null)
        </li>
    }
</ul>
```

实现系统功能导航是系统主页的任务,是系统运行的向导,是用户完成系统功能的引导地图。

7.2.6 关于我们

关于我们的主要任务是显示系统相关的信息,包括系统说明、使用单位等。关于我们的运行界面如图 7-7 所示。

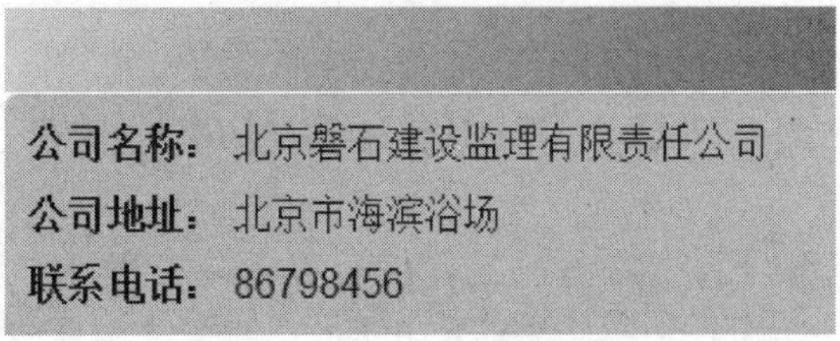

公司名称:	北京磐石建设监理有限责任公司
公司地址:	北京市海滨浴场
联系电话:	86798456

图 7-7 关于我们显示界面

关于我们显示的方法名称是"About",所属控制器是"HomeController"类,位于项目根目录下的"Controllers"目录中,其方法的代码内容如下:

```
//系统门户说明
[AllowAnonymous]
public virtual ActionResult About()
{
```

```
        ViewBag. Message ="关于公司情况说明";
        return View( );
}
```

关于我们方法对应的视图文件名称是"About. cshtml",存储于项目根目录下的"Views/Home"目录中,其代码内容如下:

```
@ {
    ViewBag. Title ="关于我们";
}
< div class ="insetText" >
    < h2 > @ ViewBag. Title < /h2 >
</div >
< div style ="border - radius:5px;background - color:#7db9e8;padding:10px;" >
    @ Html. Label("公司名称:")
    @ psjlmvc4. Properties. Resources. AppliedUnit  < br/ >
    @ Html. Label("公司地址:")
        @ psjlmvc4. Properties. Resources. AppliedUnitAddress  < br/ >
            @ Html. Label("联系电话:")
                @ psjlmvc4. Properties. Resources. AppliedTelephone  < br/ >
</div >
```

7.3 工程管理功能实现

工程管理功能编号为 A,包括工程信息编辑、工程项目调整、工程分项管理、增加新的工程、删除当前工程等子功能。功能管理实现所需要的各类文件资源位于项目根目录下的"Areas/AArea"区域目录中。

7.3.1 工程信息编辑

工程信息编辑功能需要完成对当前所选择工程的数据项目的编辑修改任务,运行界面如图 7-8 所示。

界面以两列方式并排显示当前工程的编辑修改数据项目,为了满足不同操作需求,系统同时提供了单列显示方式的编辑修改模式(在此不作说明)。工程信息编辑修改实现的方法名称是"ProjectEditDouble",所属控制器是"AProject-Controller"类,其代码内容如下:

图 7-8 工程信息编辑修改界面

```
//GET:/ProjectManaging/Edit/5
public virtual ActionResult ProjectEditDouble( AProjectList pl = null)
{
    string pcode = apps. GetSession("projectcode");
    if( String. IsNullOrEmpty( pcode))//判断工程选择与否
    {
        ViewBag. Message = psjlmvc4. Properties. Resources. ProjectNotSelected;
        return View("Error");
    }
    //模型有效
    if( ModelState. IsValid)
    {
        db. Entry( pl). State = EntityState. Modified;
        try
        {
            db. SaveChanges();
            apps. SetSession("projectname",pl. projectname);
```

7 系统功能设计与实现

```
                apps.SetSession("logdate",String.Format("{0:yyyy-MM-dd}",
                                  pl.logdate));
                ViewBag.Message ="数据存储完成....";
            }
            catch(UpdateException ee)
            {
                ViewBag.Message ="数据项目有错误,请检查...."+ ee.Message;
            }
        }
        else//模型无效
        {
            pl = db.AProjectLists.Find(pcode);
        }
        ViewBag.statusid = new SelectList(db.LStatusLists,"statusid",
                            "statusname",pl.statusid);
        ViewBag.classid = new SelectList(db.LClassLists,"classid",
                            "classname",pl.classid);
        ViewBag.levelid = new SelectList(db.LLevelLists,"levelid",
                            "levelname",pl.levelid);
        ViewBag.propertyid = new SelectList(db.LPropertyLists,
                            "propertyid","propertyname",pl.propertyid);
        ViewBag.pmanager = new SelectList(db.LUserStaffs,"userid",
                            "fullname",pl.pmanager);
        ViewBag.adjustman = new SelectList(db.LUserStaffs,"userid",
                            "fullname",pl.adjustman);
        return View(pl);
    }
```

此代码完成的主要任务有:

(1) 检查当前是否存在已选择的工程;

(2) 检查方法参数"AProjectList pl = null"是否有效;

(3) 如果有效,则存储,返回编辑视图;如果无效,检索当前记录,返回编辑视图;

(4) 检索工程所需要的相关固定数据项目的数据来源,以动态变量传递到视图,供选择使用。

方法"ProjectEditDouble"对应的视图文件名称是"ProjectEditDouble.cshtml",其代码内容如下:

```
@model psjlmvc4.Models.AProjectList
@{
    ViewBag.Title = Html.DisplayNameForModel() + "编辑修改";
}
@using(Html.BeginForm("ProjectEditDouble","AProject",new{id = Model.projectcode}))
{
    <table class="table table-condensed" style="border-bottom:2px solid #808080;margin-bottom:5px;">
        <caption><h2>@ViewBag.Title</h2></caption>
        <tr id="row00" style="background-color:aquamarine;">
            <td style="width:50%;">
                @Html.LabelFor(model => model.projectcode)
                @Html.DisplayFor(model => model.projectcode)
                @Html.HiddenFor(model => model.projectcode)
            </td>
            <td style="width:50%;">
                @Html.LabelFor(model => model.handman)
                @Html.DisplayFor(model => model.handman) ==
                @Html.DisplayFor(model => model.usHandman.fullname)
                @Html.HiddenFor(model => model.handman)
            </td>
        </tr>
        <tr id="row01">
            <td>
                @Html.LabelFor(model => model.projectname)
                @Html.EditorFor(model => model.projectname)
                @Html.ValidationMessageFor(model => model.projectname)
            </td>
            <td>
                @Html.LabelFor(model => model.paddress)
                @Html.EditorFor(model => model.paddress)
                @Html.ValidationMessageFor(model => model.paddress)
            </td>
        </tr>
        <tr id="row02">
            <td>
                @Html.LabelFor(model => model.logdate)
                @Html.EditorFor(model => model.logdate)
            </td>
            <td>
```

```
            @Html.LabelFor(model => model.pmanager)
            @Html.DropDownList("pmanager")
        </td>
    </tr>
    <tr id="row03">
        <td>
            @Html.LabelFor(model => model.statusid)
            @Html.DropDownList("statusid")
        </td>
        <td>
            @Html.LabelFor(model => model.classid)
            @Html.DropDownList("classid")
        </td>
    </tr>
    <tr id="row04">
        <td>
            @Html.LabelFor(model => model.propertyid)
            @Html.DropDownList("propertyid")
        </td>
        <td>
            @Html.LabelFor(model => model.levelid)
            @Html.DropDownList("levelid")
        </td>
    </tr>
    <tr id="row05">
        <td>
            @Html.LabelFor(model => model.begdate)
            @Html.EditorFor(model => model.begdate)
            @Html.ValidationMessageFor(model => model.begdate)
        </td>
        <td>
            @Html.LabelFor(model => model.enddate)
            @Html.EditorFor(model => model.enddate)
            @Html.ValidationMessageFor(model => model.enddate)
        </td>
    </tr>
    <tr id="row06">
        <td>
            @Html.LabelFor(model => model.pfunction)
            @Html.EditorFor(model => model.pfunction)
```

```
            @Html.ValidationMessageFor(model => model.pfunction)
        </td>
        <td>
            @Html.LabelFor(model => model.qualitytarget)
            @Html.EditorFor(model => model.qualitytarget)
            @Html.ValidationMessageFor(model => model.qualitytarget)
        </td>
    </tr>
    <tr id="row07">
        <td>
            @Html.LabelFor(model => model.contractprice)
            @Html.EditorFor(model => model.contractprice)
        </td>
        <td>
            @Html.LabelFor(model => model.contractmode)
            @Html.EditorFor(model => model.contractmode)
        </td>
    </tr>
    <tr id="row08">
        <td>
            @Html.LabelFor(model => model.planenable)
            @Html.EditorFor(model => model.planenable)
        </td>
        <td>
            @Html.LabelFor(model => model.plancode)
            @Html.EditorFor(model => model.plancode)
        </td>
    </tr>
    <tr id="row09">
        <td>
            @Html.LabelFor(model => model.buildenable)
            @Html.EditorFor(model => model.buildenable)
        </td>
        <td>
            @Html.LabelFor(model => model.buildcode)
            @Html.EditorFor(model => model.buildcode)
        </td>
    </tr>
    <tr>
        <td>
```

```
                @Html.LabelFor(model => model.planestablish)
                @Html.EditorFor(model => model.planestablish)
            </td>
            <td>
                @Html.LabelFor(model => model.establishcode)
                @Html.EditorFor(model => model.establishcode)
            </td>
        </tr>
        <tr>
            <td>
                @Html.LabelFor(model => model.investproperty)
                @Html.EditorFor(model => model.investproperty)
            </td>
            <td>
                @Html.LabelFor(model => model.investbody)
                @Html.EditorFor(model => model.investbody)
            </td>
        </tr>
        <tr>
            <td>
                @Html.LabelFor(model => model.builduparea)
                @Html.EditorFor(model => model.builduparea)
            </td>
            <td>
                @Html.LabelFor(model => model.reportarea)
                @Html.EditorFor(model => model.reportarea)
            </td>
        </tr>
        <tr>
            <td>
                @Html.LabelFor(model => model.supervisionprice)
                @Html.EditorFor(model => model.supervisionprice)
            </td>
            <td>
                @Html.LabelFor(model => model.adjustman)
                @Html.DropDownList("adjustman")
            </td>
        </tr>
        <tr>
            <td>
```

```
                @Html.LabelFor(model => model.adjustdate)
                @Html.EditorFor(model => model.adjustdate)
            </td>
            <td>
                @Html.LabelFor(model => model.begpicket)
                @Html.EditorFor(model => model.begpicket)
                @Html.ValidationMessageFor(model => model.begpicket)
            </td>
        </tr>
        <tr>
            <td>
                @Html.LabelFor(model => model.endpicket)
                @Html.EditorFor(model => model.endpicket)
                @Html.ValidationMessageFor(model => model.endpicket)
            </td>
            <td>
                @Html.LabelFor(model => model.remark)
                @Html.EditorFor(model => model.remark)
            </td>
        </tr>
    </table>
    <div class="hidden-print">
        <button type="submit" class="btn btn-primary glyphicon glyphicon-save">存储</button>
        @Html.ActionLink("返回","Index","Home",new{Area=""},new{@class="btn btn-primary glyphicon glyphicon-home"})
        <span class="message">@ViewBag.Message</span>
    </div>
}
@section Scripts{
    @Scripts.Render("~/bundles/jqueryval")
}
<script type="text/javascript">
    $("#logdate").datepicker({dateFormat:"yy-mm-dd",changeMonth:true,changeYear:true});
    $("#begdate").datepicker({dateFormat:"yy-mm-dd",changeMonth:true,changeYear:true});
    $("#enddate").datepicker({dateFormat:"yy-mm-dd",changeMonth:true,changeYear:true});
    $("#adjustdate").datepicker({dateFormat:"yy-mm-dd",changeMonth:true,changeYear:true});
</script>
```

数据项目的排列使用了表格标签（Table），JavaScript代码段实现了日期输入选择器控件与日期输入控件的绑定及自动切换功能。

7.3.2 工程项目调整

工程项目调整功能完成工程项目调整数据记录的检索、显示、新增、编辑、删除等任务，所需要的方法定义在"AProjectAdjustController"控制器中，其中"Index"是记录检索和功能操作入口的方法，代码内容如下：

```
//GET:/AArea/AProjectAdjust/
public ActionResult Index()
{
    string pcode = apps.GetSession("projectcode");
    if(string.IsNullOrEmpty(pcode))
    {
        ViewBag.Message ="没有选择工程！请使用[工程选择]功能选择要处理的工程…";
        return View("Error");
    }
    ViewBag.ModelDisplayName =
        apps.GetModelDisplayName(typeof(AProjectAdjust));
    return View(db.AProjectAdjusts.Where(a => a.projectcode == pcode).ToList());
}
```

首先判断当前工程是否存在，不存在则显示相应提示；如果存在则检索此工程的所有项目调整记录，并传递到对应的视图进行显示。"Index"方法对应的视图文件名称是"Index.cshtml"，其代码内容如下：

```
@model IEnumerable<psjlmvc4.Models.AProjectAdjust>
@{
    ViewBag.Title = ViewBag.ModelDisplayName +"记录数据列表(Index For Record Data)";
}
<script type="text/javascript">
    $(document).ready(function(){
        var oldback = $("#dtable tr").css("background-color");
        var oldcolor = $("#dtable tr").css("color");
        $("#dtable tr td").click(function(){
            $("#dtable tr").css("background-color",oldback);
            $("#dtable tr").css("color",oldcolor);
            $(this).parent().css("background-color","#03a5d1");
            $(this).parent().css("color","white");
```

```html
        });
    })
</script>
<table id="dtable" class="table-striped">
    <caption><h2>@ViewBag.Title</h2></caption>
    <thead>
    <tr>
        <th>
            @Html.DisplayNameFor(model => model.id)
        </th>
        <th>
            @Html.DisplayNameFor(model => model.adjustcontent)
        </th>
        <th>
            @Html.DisplayNameFor(model => model.adjustarea)
        </th>
        <th>
            @Html.DisplayNameFor(model => model.handman)
        </th>
        <th>
            @Html.DisplayNameFor(model => model.adjustdate)
        </th>
        <th>
            @Html.DisplayNameFor(model => model.remark)
        </th>
        <th class="noprint" style="width:150px;"></th>
    </tr>
    </thead>
    <tbody>
@foreach(var item in Model){
    <tr>
        <td>
            @Html.DisplayFor(modelItem => item.id)
        </td>
        <td>
            @Html.DisplayFor(modelItem => item.adjustcontent)
        </td>
        <td>
            @Html.DisplayFor(modelItem => item.adjustarea)
        </td>
        <td>
```

```
            @ Html. DisplayFor( modelItem = > item. handman)
        </td >
        < td >
            @ Html. DisplayFor( modelItem = > item. adjustdate)
        </td >
        < td >
            @ Html. DisplayFor( modelItem = > item. remark)
        </td >
        < td class ="noprint" >
            @ Html. ActionLink("编辑","Edit",new{ id = item. id } ) |
            @ Html. ActionLink("详细","Details",new{ id = item. id } ) |
            @ Html. ActionLink("删除","Delete",new{ id = item. id } )
        </td >
    </tr >
}
< tbody >
</table >
< p class ="noprint" >
    @ Html. ActionLink ("新增","Create", null, new { @ class ="btn btn - primary glyphicon
glyphicon - plus"} )
    < button id ="print"onclick ="window. print( ); return false;"class ="btn btn - primary
glyphicon glyphicon - print" > 打印 </button >
    @ Html. ActionLink("返回","Index","Home", new { Area =""} , new { @ class ="btn btn -
primary glyphicon glyphicon - home"} )
    @ ViewBag. saveid
</p >
```

"Index. cshtml" 运行的效果界面如图 7 -9 所示。

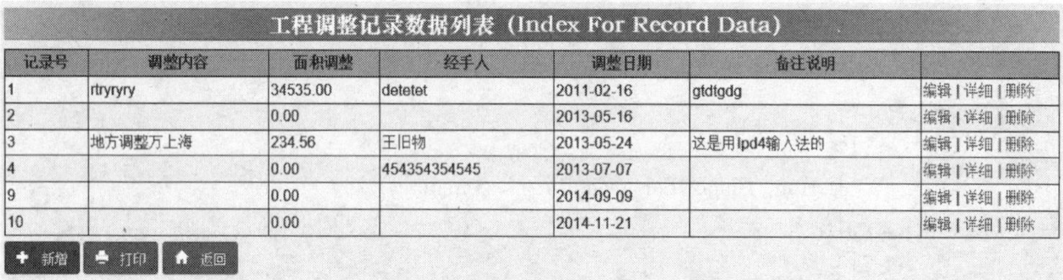

工程调整记录数据列表 (Index For Record Data)

记录号	调整内容	面积调整	经手人	调整日期	备注说明	
1	rtryryry	34535.00	detetet	2011-02-16	gtdtgdg	编辑 | 详细 | 删除
2		0.00		2013-05-16		编辑 | 详细 | 删除
3	地方调整万上海	234.56	王旧物	2013-05-24	这是用lpd4输入法的	编辑 | 详细 | 删除
4		0.00	454354354545	2013-07-07		编辑 | 详细 | 删除
9		0.00		2014-09-09		编辑 | 详细 | 删除
10		0.00		2014-11-21		编辑 | 详细 | 删除

图 7 -9 工程项目调整视图运行界面

界面中与显示记录无关的操作链接有"新增"和"返回",与记录管理有关的操作链接有"编辑"、"详细"和"删除",这些都是由 CRUD 模板生成的管理代码。在第 6 章中有关 CRUD 模板的使用已做了详细介绍,此处和后续章节涉及 CRUD 常规操作的内容将不再赘述。

7.3.3 工程分项管理

工程分项管理功能完成工程项目分解（子工程）数据记录的检索、显示、新增、编辑、删除等任务，所需要的方法定义在"AProjectPartController"控制器中，其中"Index"是记录检索和功能操作入口的方法，代码内容如下：

```
//GET:/AArea/AProjectPart/
public ActionResult Index()
{
    string pcode = apps.GetSession("projectcode");
    if(string.IsNullOrEmpty(pcode))
    {
        ViewBag.Message = "没有选择工程！请使用[工程选择]功能选择要处理的工程...";
        return View("Error");
    }
    ViewBag.ModelDisplayName =
        apps.GetModelDisplayName(typeof(AProjectPart));
    return View(db.AProjectParts.Where(a => a.projectcode == pcode).ToList());
}
```

首先判断当前工程是否存在，不存在则显示相应提示；如果存在则检索此工程的所有项目调整记录，并传递到对应的视图进行显示。"Index"方法对应的视图文件名称是"Index.cshtml"，其代码内容如下：

```
@model IEnumerable<psjlmvc4.Models.AProjectPart>
@{
    ViewBag.Title = ViewBag.ModelDisplayName + "记录数据列表(Index For Record Data)";
}
<script type="text/javascript">
    $(document).ready(function() {
        var oldback = $("#dtable tr").css("background-color");
        var oldcolor = $("#dtable tr").css("color");
        $("#dtable tr td").click(function() {
            $("#dtable tr").css("background-color", oldback);
            $("#dtable tr").css("color", oldcolor);
            $(this).parent().css("background-color", "#03a5d1");
            $(this).parent().css("color", "white");
```

```
            });
        })
</script>
<table id="dtable">
    <caption><h2>@ViewBag.Title</h2></caption>
    <thead>
    <tr>
        <th>
            @Html.DisplayNameFor(model => model.AProjectPartID)
        </th>
        <th>
            @Html.DisplayNameFor(model => model.projectcode)
        </th>
        <th>
            @Html.DisplayNameFor(model => model.partname)
        </th>
        <th>
            @Html.DisplayNameFor(model => model.buildingarea)
        </th>
        <th class="noprint" style="width:150px;"></th>
    </tr>
    </thead>
    <tbody>
@foreach(var item in Model){
    <tr>
        <td>
            @Html.DisplayFor(modelItem => item.AProjectPartID)
        </td>
        <td>
            @Html.DisplayFor(modelItem => item.projectcode)
        </td>
        <td>
            @Html.DisplayFor(modelItem => item.partname)
        </td>
        <td>
            @Html.DisplayFor(modelItem => item.buildingarea)
        </td>
        <td class="noprint">
            @Html.ActionLink("编辑","Edit",new{id=item.AProjectPartID}) |
            @Html.ActionLink("详细","Details",new{id=item.AProjectPartID}) |
            @Html.ActionLink("删除","Delete",new{id=item.AProjectPartID})
```

```
            </td>
        </tr>
    }
    <tbody>
    </table>
    <p class="noprint">
        @Html.ActionLink("新增","Create",null,new{@class="btn btn-primary glyphicon glyphicon-plus"})
        <button id="print"onclick="window.print();return false;"class="btn btn-primary glyphicon glyphicon-print">打印</button>
        @Html.ActionLink("返回","Index","Home",new{Area=""},new{@class="btn btn-primary glyphicon glyphicon-home"})
    </p>
```

“Index. cshtml”运行的效果界面如图 7－10 所示。

工程分项记录数据列表（Index For Record Data）

记录号	工程编号	工程分项名称	建筑面积(m2)	
1003	2013-5001	屋顶防水	12345.77	编辑 | 详细 | 删除
2003	2013-5001		0.00	编辑 | 详细 | 删除
2004	2013-5001		0.00	编辑 | 详细 | 删除
2005	2013-5001		0.00	编辑 | 详细 | 删除
2010	2013-5001		0.00	编辑 | 详细 | 删除

新增　打印　返回

图 7－10　工程分项管理视图运行界面

界面中与显示记录无关的操作链接有“新增”和“返回”，与记录管理有关的操作链接有“编辑”、“详细”和“删除”。

7.3.4　增加新的工程

增加新的工程实现的方法名称是“ProjectAdd”，所属控制器是项目区域目录“Controllers”下的“AProjectController. cs”类，具体代码内容如下：

```
//GET:/AArea/Project/ProjectAdd
public ActionResult ProjectAdd()
{
    AProjectAdd apm = new AProjectAdd();
    apm.handman = apps.GetSession("userid");
    apm.logdate = DateTime.Today;
```

```
        ViewBag. Message ="增加新的工程....";
        ViewBag. fullname = apps. GetSession("fullname");
        return View(apm);
}
//POST:/AArea/AProject/ProjectAdd
[HttpPost]
public ActionResult ProjectAdd(AProjectAdd apm)
{
        ViewBag. Message ="编号为的工程数据存储完成....";
        if(ModelState. IsValid)
        {
                AProjectList pl = new AProjectList();
                pl. projectcode = apm. projectcode;
                pl. projectname = apm. projectname;
                pl. paddress = apm. paddress;
                pl. handman = apm. handman;
                pl. logdate = apm. logdate;
                ViewBag. fullname = apps. GetSession("fullname");
                db. AProjectLists. Add(pl);
                try
                {
                        db. SaveChanges();
                        ViewBag. Message ="编号为["+ pl. projectcode +"]的工程数据存储完成....";
                }
                catch
                {
                        ViewBag. Message ="编号为["+ pl. projectcode +"]的数据已经存在!";
                }
        }
        return View(apm);
}
```

增加新的工程方法"ProjectAdd"有两个状态：一个是没有参数的调用视图的入口方法；另一个是通过"HttpPost"方式返回视图数据的方法，实体模型"ProjectAdd"是传递和暂存视图数据的临时模型。在第一个状态中，定义模型实例，确定相关项目值，传递给视图；在第二个状态中，接受视图输入数据，个性工程实体模型"AProjectList"，将数据存入数据库，这里处理的有关工程实体的数据项目主要有工程编号、工程名称、注册日期、经手人，其他数据项目通过工程信息编辑进行修改并完善，可以由具体施工单位修改完善。

方法"ProjectAdd"对应的视图文件名称是"ProjectAdd.cshtml"，其具体代码内容如下：

```
@model psjlmvc4.Models.AProjectAdd
@{
    ViewBag.Title = "工程管理-增加新工程(Create New Recoed)";
}
@using(Html.BeginForm())
{
    @Html.ValidationSummary(true)
    <table id="tt-table">
        <caption><h2>@ViewBag.Title</h2></caption>
        <tr>
            <td class="td-label">
                @Html.LabelFor(model => model.projectcode)
            </td>
            <td class="td-field">
                @Html.EditorFor(model => model.projectcode)
                @Html.ValidationMessageFor(model => model.projectcode)
            </td>
        </tr>
        <tr>
            <td class="td-label">
                @Html.LabelFor(model => model.projectname)
            </td>
            <td class="td-field">
                @Html.EditorFor(model => model.projectname)
                @Html.ValidationMessageFor(model => model.projectname)
            </td>
        </tr>
        <tr>
            <td class="td-label">
                @Html.LabelFor(model => model.paddress)
            </td>
            <td class="td-field">
                @Html.EditorFor(model => model.paddress)
                @Html.ValidationMessageFor(model => model.paddress)
            </td>
        </tr>
```

```
            <tr>
                <td class="td-label">
                    @Html.LabelFor(model=>model.logdate)
                </td>
                <td class="td-field">
                    @Html.EditorFor(model=>model.logdate)
                    @Html.ValidationMessageFor(model=>model.logdate)
                </td>
            </tr>
            <tr>
                <td class="td-label">
                    @Html.LabelFor(model=>model.handman)
                </td>
                <td class="td-field">
                    @Html.HiddenFor(model=>model.handman)
                    @Html.DisplayFor(model=>model.handman)==
                    @Html.Encode(ViewBag.fullname)
                </td>
            </tr>
            <tr style="border-top:1px solid black;">
                <td class="td-label">操作说明</td>
                <td class="td-field" style="padding:10px;font-size:large;">
                    <ol>
                        <li>
                            通过输入"工程编号、工程名称和注册日期",新增加工程,有关工程的其他数据项目通过[工程编辑]功能修改;
                        </li>
                        <li>
                            "工程编号"字段内容必须输入,数据格式为[××××-××××];其中前四位(××××)代表年份,后四位(××××)代表序号。
                        </li>
                        <li>
                            "经手人员"内容为只读,不能输入或修改。
                        </li>
                    </ol>
                </td>
            </tr>
        </table>
        <p class="hidden-print">
            <button type="submit" class="btn btn-primary glyphicon glyphicon-save">存储</button>
```

```
            @Html.ActionLink("返回","Index","Home",new{Area=""},new{@class="btn
btn-primary glyphicon glyphicon-home"})
            <span class="message">@ViewBag.Message</span>
        </p>
}
@section Scripts{
    @Scripts.Render("~/bundles/jqueryval")
}
<script type="text/javascript">
    $("#logdate").datepicker({dateFormat:"yy-mm-dd",changeMonth:true,changeYear:true});
</script>
```

视图运行的界面如图7-11所示。

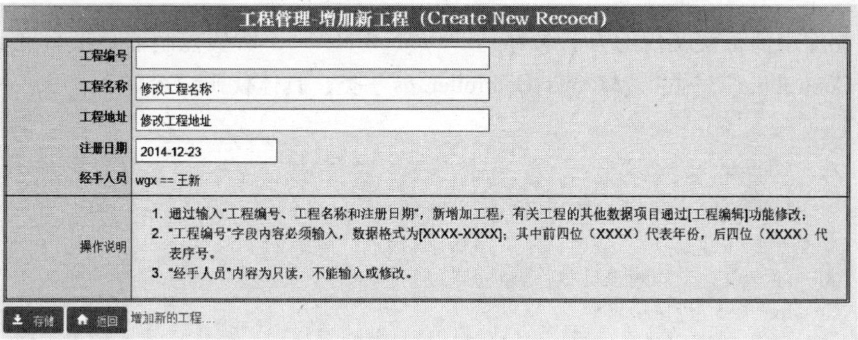

图7-11 增加新的工程视图运行界面

界面上对于工程数据项目有关的输入内容的要求有详细的说明，特别是有关工程编号的设计和输入，还提供远程模型级别的重复验证方法"CodeValidate"，以阻止重复工程编号的存在。"CodeValidate"是系统中模型级别编号重复验证的通用方法名称，存在于需要编号重复验证的对应实体模型管理的控制器类中。"CodeValidate"实现的具体代码内容如下：

```
//验证代码是否存在
[HttpGet]
public virtual ActionResult CodeValidate(string projectcode=null)
{
    bool exists = db.AProjectLists.Find(projectcode) == null;
    return Json(exists,JsonRequestBehavior.AllowGet);
}
```

在实体模型中的使用形式代码内容如下所示：

```
[Key]
[Remote("CodeValidate","AProject",ErrorMessage ="所输入的工程编号已存在")]
    [Display(Name ="工程编号")]
    [Required(ErrorMessage ="此字段内容为必须填写")]
    [StringLength(50)]
    public string projectcode{get;set;}
```

这里的"Remote"就是调用远程验证方法，要实现此方法，在视图中需要引入两个 JS 文件：一个是"jquery. validate. js"，另一个是"jquery. validate. unobtrusive. js"或"jquery. validate. unobtrusive. min. js"，后文同理。

7.3.5　删除当前工程

删除当前工程实现的方法名称是"ProjectDelete"，所属控制器是项目区域目录"Controllers"下的"AProjectController. cs"类，具体代码内容如下：

```
/// <summary>
///Action Name:ProjectDelete,HttpGet openned
/// </summary>
/// <param> </param>
/// <returns> </returns>
public virtual ActionResult ProjectDelete()
{
    string pcode = apps. GetSession("projectcode");
    if(String. IsNullOrEmpty(pcode))
    {
        ViewBag. Message ="没有选择工程！请使用[工程选择]
                          选择要删除的工程...";
        return View("Error");
    }
    else
    {
        ViewBag. Message ="请选择[确认删除选项]，
                          并点击[确认删除]以完成删除工程...";
        return View();
    }
}
```

```
//GET:/Project/ProjectDelete/5
[HttpPost]
public virtual ActionResult ProjectDelete(bool yesno = false, string pcode = "")
{
    if(yesno = = false)
    {
        ViewBag. Message = "请首先选择[确认删除选项],
                        然后点击[确认删除],删除工程...";
    }
    else
    {
        AProjectList rd = db. AProjectLists. Find(pcode);
        db. AProjectLists. Remove(rd);
        db. SaveChanges();
        apps. SetSession("projectcode","");
        apps. SetSession("projectname","");
        apps. SetSession("logdate","");
        ViewBag. Message = "编号为[" + pcode + "]的工程删除完成!";
    }
    return View();
}
```

删除当前工程方法"ProjectDelete"有两个状态：一个是没有参数的视图调用入口方法；另一个是通过"HttpPost"方式接受视图数据的方法。其中接受的参数一个是删除确认变量"yesno"，另一个是工程编号"pcode"。在第一个状态中，通过当前工程的编号，判断是否有此工程存在，如不存在，显示运行状态提示信息；如存在，显示工程删除的确认视图。

删除当前工程方法"ProjectDelete"对应的视图文件名称是"ProjectAdd. cshtml"，其具体代码内容如下：

```
@ using psjlmvc4. Models
@ {
    ViewBag. Title = "工程管理 – 删除当前工程(Delete a Record)";
    AppService apps = new AppService();
    var pcode = apps. GetSession("projectcode");
    var pname = apps. GetSession("projectname");
    var pdate = apps. GetSession("logdate");
}
```

```
<div class = "panel panel-primary"style = "width:70%;margin:auto;">
    <div class = "panel-heading">
        @ViewBag.Title
    </div>
    <div class = "panel-body">
        @Html.Label("工程编号:")
        @Html.Label(pcode)
        <br/>
        @Html.Label("工程名称:")
        @Html.Label(pname)
        <br/>
        @Html.Label("注册日期:")
        @Html.Label(pdate)
        <hr/>
        @using(Html.BeginForm("ProjectDelete","AProject",FormMethod.Post,new{@class = "form-horizontal",role = "form"}))
        {
            @Html.Hidden("pcode",pcode)
            <div class = "form-group">
                <div class = "col-sm-offset-2 col-sm-10">
                    <label>
                        <input type = "checkbox"value = "N"title = "确认删除选项"/>确认删除选项
                    </label>
                </div>
            </div>
            <div class = "form-group">
                <div class = "col-sm-offset-2 col-sm-10">
                    <button type = "submit"class = "btn btn-primary glyphicon glyphicon-remove">确认删除</button>
                    @Html.ActionLink("返回","Index","Home",new{Area = ""},new{@class = "btn btn-primary glyphicon glyphicon-home"})
                </div>
            </div>
        }
    </div>
    <div class = "panel-footer">
        <span class = "message">@ViewBag.Message</span>
    </div>
</div>
```

视图运行的界面如图 7-12 所示。

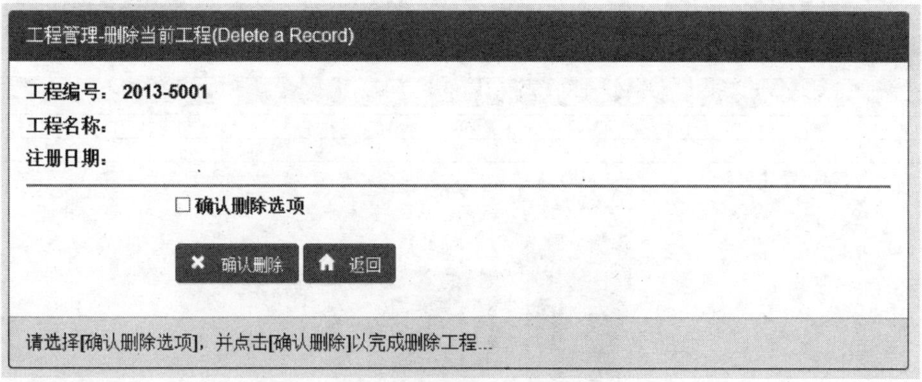

图 7-12　删除当前工程视图运行界面

有关工程信息内容的显示是通过 Session 变量实现的。

7.4　其他功能实现

在 CRUD 模板的支持下，多数实体模型的管理实现方法相似，为节省篇幅，在众多功能中特别选择非 CRUD 模板框架的功能进行说明，并介绍其实现方法和过程。

本节介绍的其他功能包括"J-查询统计"、"K-系统管理"、"L-基础数据"、"O-其他辅助信息管理"等。

7.4.1　J-查询统计

查询统计功能模块主要包括与工程管理统计有关的功能实现，目前所包含的子功能有"工程项目统计"、"管理表格统计"、"信息回复统计"、"监理工作统计（D9）"、"年工程量分类统计"、"工程款审批及支付汇总表（D8）"等。功能模块代码为"J"，其所有功能实现所需要的文件资源在区域"Area/JArea"中。现以"工程项目统计"功能为例说明其要求和实现方法。

7.4.1.1　统计要求

工程项目统计主要完成不同时期或时间段内的工程数量统计任务，并按照年、月、日分类；功能要求以"月"为单位进行数量上的比较（同比和环比），结果同时以表格和图表形式显示。

工程项目统计的视图文件名称是"Index.cshtml"，其运行界面如图 7-13 所示。

图7-13　工程项目统计视图运行界面

"Index.cshtml"文件的代码内容如下：

```
@model psjlmvc4.Models.JCondition
@{
    ViewBag.Title ="工程量统计";
}
<h2>@ViewBag.Title</h2>
@using(Html.BeginForm("Index","ProjectStatistics",new{@class ="form-horizontal",role ="form"}))
{
    <table style ="width:100%;border:none;">
        <tr style ="border:none;">
            <td style ="border:none;">
                检索条件：
            </td>
            <td style ="border:none;">
                @Html.LabelFor(model => model.begDate)
                @Html.TextBoxFor(model => model.begDate,Model.begDate.ToString("yyyy-MM-dd"),new{style ="width:120px;"})
            </td>
            <td style ="border:none;">
                @Html.LabelFor(model => model.endDate)
```

```
                @Html.TextBoxFor(model => model.endDate, Model.begDate.ToString
("yyyy-MM-dd"), new { style = "width:120px;"})
            </td>
            <td style = "border:none;">
                @Html.LabelFor(model => model.chartType)
                @Html.DropDownListFor(model => model.chartType, ViewBag.charttype
as SelectList, new { style = "width:120px;"} )
            </td>
            <td style = "border:none;">
                <button type = "submit" class = "btn btn-primary glyphicon glyphicon-
search">确认</button>
                @Html.ActionLink("返回", "Index", "Home", new { Area = ""}, new { @
class = "btn btn-primary glyphicon glyphicon-home"} )
            </td>
        </tr>
    </table>
}
<hr/>
<ul class = "nav nav-pills nav-tabs-justified">
    <li class = "active"><a href = "#tab-year" data-toggle = "tab">年工程量</a></
li>
    <li><a href = "#tab-month" data-toggle = "tab">月工程量</a></li>
    <li><a href = "#tab-day" data-toggle = "tab">日工程量</a></li>
    <li><a href = "#tab-compare" data-toggle = "tab">同期比较</a></li>
</ul>
<div class = "tab-content" style = "text-align:center;">
    <div class = "tab-pane active" id = "tab-year">
        @{Html.RenderAction("YearCountPartial", new { date1 = Model.begDate, date2 =
Model.endDate } );}
        <img id = "ychart" alt = "ychart" src = "@Url.Action("YearChart", new
                                              {
                                                  date1 = Model.begDate,
                                                  date2 = Model.endDate,
                                                  ctype = Model.chartType
                                              })"/>
    </div>
    <div class = "tab-pane" id = "tab-month">
        @{Html.RenderAction("MonthCountPartial", new { date1 = Model.begDate } );}
        <img id = "mchart" alt = "Mchart" src = "@Url.Action("MonthChart", new { date1 =
Model.begDate, ctype = Model.chartType })"/>
```

```
            </div>
            <div class="tab-pane" id="tab-compare">
                @{Html.RenderAction("MonthCountPartial",new{date1=Model.begDate});}
                @{Html.RenderAction("MonthCountPartial",new{date1=Model.begDate.AddYears(-1)});}
                <img id="mchart" alt="Mchart" src="@Url.Action("MonthCompareChart",new{date1=Model.begDate,ctype=Model.chartType})"/>
            </div>
            <div class="tab-pane" id="tab-day">
                @{Html.RenderAction("DayCountPartial",new{date1=Model.begDate});}
                <img id="mchart" alt="Mchart" src="@Url.Action("DayCountChart",new{date1=Model.begDate,ctype=Model.chartType})"/>
            </div>
        </div>
    <script type="text/javascript">
        $("#begDate").datepicker();
        $("#endDate").datepicker();
    </script>
```

7.4.1.2 Index 方法

实现视图的方法名称是"Index",定义在名称为"ProjectStatistics"的控制器类中,具体代码内容如下:

```
public ActionResult Index(JCondition jc)
{
    if(jc == null)
    {
        jc = new JCondition();
    }
    ViewBag.charttype = apps.ChartTypeList();
    return View(jc);
}
```

在方法中,关键是利用实体模型"JCondition"存储所有的检索条件变量,然后通过分部视图方法接受参数,以分部视图方式完成所要求的统计任务。

7.4.1.3 年工程量统计分部视图和实现方法

年工程量统计分部视图的方法名称是"YearCountPartial",具体代码内容如下:

```
public ActionResult YearCountPartial(DateTime date1, DateTime date2)
{
    int y1 = date1.Year, y2 = date2.Year;
    var sl = new StatisticsList(date1, date2);
    var rd = db.AProjectLists.Where(p => p.logdate.Year >= y1 &&
            p.logdate.Year <= y2);
    for(int i = y1; i <= y2; i++)
    {
        sl.xValue.Add(i.ToString());
        sl.yValue.Add(rd.Count(p => p.logdate.Year == i));
    }
    ViewBag.Message = y1 + "-" + y2 + "年工程量统计";
    return PartialView(sl);
}
```

该方法接受两个参数：date1——统计开始日期；date2——统计结束日期，统计结果存储于实体模型"StatisticsList"中，并通过"return PartialView(sl);"传递分部视图进行显示。分部视图"YearCountPartial.cshtml"的实现代码内容如下：

```
@model psjlmvc4.Models.StatisticsList
@{
    ViewBag.Title = ViewBag.Message;
}
<table id="dtable">
    <caption><h2>@ViewBag.Title</h2></caption>
    <tr>
        <td>年份</td>
        @foreach(var item in Model.xValue)
        {
            <td>@item</td>
        }
        <td>合计</td>
    </tr>
    <tr>
        <td>数量</td>
        @foreach(var item in Model.yValue)
        {
            <td>@item</td>
```

```
            <td>@Model.yValue.Sum()</td>
        </tr>
</table>
```

相应的图表显示实现的方法名称是"YearChart",代码内容如下:

```
//年份工程量统计图表显示
public ActionResult YearChart(DateTime date1, DateTime date2,
            string ctype = "Column")
{
    int y1 = date1.Year, y2 = date2.Year;
    var sl = new StatisticsList(date1, date2);
    var rd = db.AProjectLists.Where(p = > p.logdate.Year > = y1 &&
            p.logdate.Year < = y2);
    for(int i = y1; i < = y2; i + +)
    {
        sl.xValue.Add(i.ToString());
        sl.yValue.Add(rd.Count(p = > p.logdate.Year = = i));
    }
    ViewBag.Message = y1 + " - " + y2 + "年工程量统计";
    var key = new Chart(width: 1100, height: 400, theme: ChartTheme.Green);
    key.AddTitle(ViewBag.Message);
    //key.AddLegend();只有一个系列,可以不加图例
    key.AddSeries(name: "yearvalue1", chartType: ctype,
            xValue: sl.xValue, yValues: sl.yValue);
    var filePathName = Server.MapPath(" ~ /Images/") + "chart02.jpg";
    FileInfo df = new FileInfo(filePathName);
    df.Delete();
    key.Save(path: filePathName);
    byte[ ] bt = key.GetBytes();
    var tp = key.GetType().ToString();
    return new FileContentResult(bt, tp);
}
```

该方法以"return new FileContentResult(bt, tp);"形式返回图形文件,在视图中进行显示。

7.4.2 K - 系统管理

系统管理功能完成系统设置有关的任务。系统管理功能包括系统角色管理、

系统功能管理、角色功能管理、重置用户密码、操作日志管理等子功能。功能实现的文件资源存储于区域"Areas/KArea"中。现以"角色功能"管理为例说明其实现的方法和过程。

角色功能管理的任务是定义系统角色便拥有的系统功能,从而实现部分操作权限的控制,功能实现的所有方法定义在控制器"KGroupFunController"类中,其中"Index"是功能入口方法,具体代码内容如下:

```
//GET:/KArea/KGroupFun/
public virtual ActionResult Index()
{
    return View(db.KGroupLists.ToList());
}
```

方法通过"return View(db.KGroupLists.ToList());"形式,传递系统角色目录到视图进行显示。"Index"方法对应的视图文件名称是"Index.cshtml",具体代码内容如下:

```
@model IEnumerable<psjlmvc4.Models.KGroupList>
@{
    ViewBag.Title = "系统角色功能设置";
}
<h2>@ViewBag.Title</h2>
<div class="row">
    <div class="col-sm-5">
        <table id="dtable">
            <caption><h2>系统角色</h2></caption>
            <tr>
                <th>@Html.DisplayNameFor(model => model.groupid)</th>
                <th>@Html.DisplayNameFor(model => model.groupname)</th>
                <th>@Html.DisplayNameFor(model => model.remark)</th>
                <th></th>
            </tr>
            @foreach(var item in Model)
            {
                <tr>
                    <td>@Html.DisplayFor(modelitem => item.groupid)</td>
                    <td>@Html.DisplayFor(modelitem => item.groupname)</td>
                    <td>@Html.DisplayFor(modelitem => item.remark)</td>
                    <td>
```

```
                    @ Ajax. ActionLink("功能设置","FunListPartial","KGroupFun",
new{id = item. groupid,gname = item. groupname},
                    new AjaxOptions{UpdateTargetId = "div - funlist"})
                </td>
            </tr>
        }
    </table>
    <p>
        @ Html. ActionLink("返回","Index","Home",new{Area = ""},new{@ class = "
btn btn - primary glyphicon glyphicon - home"})
    </p>
</div>
<div class = "col - sm - 7">
    <div id = "div - funlist">
        功能记录显示
    </div>
</div>
</div>
```

视图运行后的界面如图 7 - 14 所示。

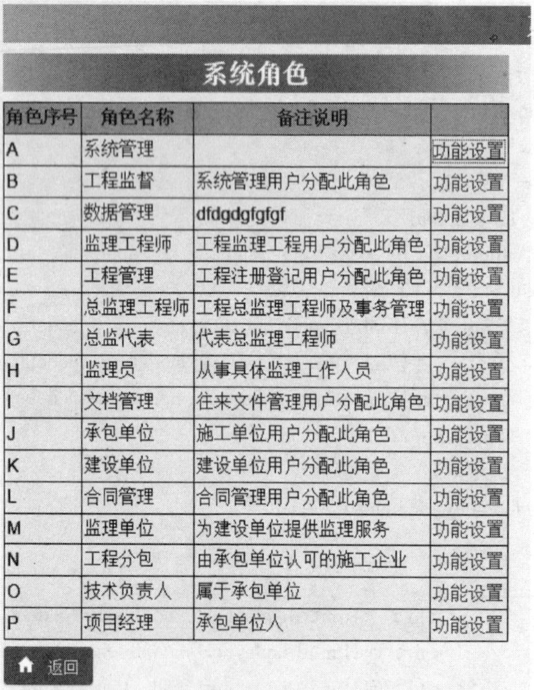
系统角色

角色序号	角色名称	备注说明	
A	系统管理		功能设置
B	工程监督	系统管理用户分配此角色	功能设置
C	数据管理	dfdgdgfgfgf	功能设置
D	监理工程师	工程监理工程用户分配此角色	功能设置
E	工程管理	工程注册登记用户分配此角色	功能设置
F	总监理工程师	工程总监理工程师及事务管理	功能设置
G	总监代表	代表总监理工程师	功能设置
H	监理员	从事具体监理工作人员	功能设置
I	文档管理	往来文件管理用户分配此角色	功能设置
J	承包单位	施工单位用户分配此角色	功能设置
K	建设单位	建设单位用户分配此角色	功能设置
L	合同管理	合同管理用户分配此角色	功能设置
M	监理单位	为建设单位提供监理服务	功能设置
N	工程分包	由承包单位认可的施工企业	功能设置
O	技术负责人	属于承包单位	功能设置
P	项目经理	承包单位人	功能设置

图 7 - 14 系统角色功能视图运行界面

通过"功能设置"链接可以实现对应角色所拥有系统功能的设置。"功能设置"实现的方法名称是"FunListPartial",其代码内容如下:

```csharp
///角色功能设置-显示系统功能并选择分部视图
public virtual ActionResult FunListPartial(string id,string gname)
{
    TempData["groupid"] = id;//id:角色序号
    ViewBag.gname = gname;
    var rdset1 = (from g in db.KGroupFuns
                  where g.groupid == id select g).ToList();
    var rdset2 = db.KFunLists.ToList();
    int cn = 0;
    foreach(var item in rdset2)
    {
        var rd = from r in rdset1 where r.funcode == item.funcode select r;
        if(rd.Count() > 0)
        {
            item.yesno = true;
        }
        else
        {
            item.yesno = false;
        }
        cn++;
        UpdateModel(item);
    }
    ViewBag.cn = cn;
    return PartialView(rdset2);
}
///角色功能设置-将选择后的功能存储到数据库
[HttpPost]
public virtual ActionResult SetGroupFun(string id,string[] yesno)
{
    try
    {
        ViewBag.groupid = id;
        var rd = from g in db.KGroupFuns where g.groupid == id select g;
        int dels = 0;
        foreach(var item in rd.ToList())
        {
```

```
                db. KGroupFuns. Remove( item );
                dels + + ;
            }
            db. SaveChanges( );
            int adds = 0;
            foreach( var item in yesno)
            {
                KGroupFun gf = new KGroupFun( );
                gf. groupid = id;
                gf. funcode = item;
                db. KGroupFuns. Add( gf );
                adds + + ;
            }
            db. SaveChanges( );
            ViewBag. message = "删除功能数:" + dels + ";增加功能数:" + adds;
        }
        catch( Exception ee)
        {
            ViewBag. message = ee. Message + "角色功能设置失败!";
        }
        return JavaScript( "alert( '" + ViewBag. message + "' );" );
    }
```

"FunListPartial"方法完成给定角色（参数ID）目前功能状态的显示，相应的视图文件名称是"FunListPartial. cshtml"；"SetGroupFun"方法完成给定角色功能设置后的数据存储任务，并以"JavaScript"方法返回处理结果，没有对应的视图文件。分部视图的显示界面如图7-15所示。

7.4.3　L-基础数据

　　基础数据功能完成对系统所需要的基础数据的管理任务。基础数据管理功能包括往来单位管理、用户职员管理、单位部门管理、监理设备管理、工程状态管理、工程类别管理、工程性质管理、工程级别管理、单位类型管理、工作任务管理、报表目录管理、文章类别管理、新闻公告管理、合同类别管理、工程合同管理等子功能。实现功能所使用的文件资源存储于区域"Areas/LArea"中，并基本都可以用CRUD框架实现。现以"往来单位"子功能为例说明其实现的方法和过程。

7.4 其他功能实现

"系统管理" 角色功能设置			
	编号	功能名称	功能说明
☑	A	A-工程管理	有关工程基础信息建设管理
☑	A01	工程信息编辑	编辑修改当前工程数据项目
☑	A03	工程项目调整	工程项目调整记录表
☑	A04	表格方式编辑	表格方式数据编辑
…	…	…	…
☑	M	M-安全管理	
☑	M01	施工资质审核	审核施工单位相关施工资质
☑	M02	施工方案审核	施工方案审核
☑	M03	安全管理交底	制订安全管理细则
☑	M04	工作联系单-CA	工作联系单（C1）
☑	M05	工程暂停令-BD	工程暂停令（B4）
☑	M06	安全问题事故处理	安全问题事故处理
☑	O	O-其他辅助信息管理	其他辅助信息管理
☑	O01	发言内容类别	发言内容类别管理
☑	O02	用户发言	用户发言
☑	P	P-施工单位	施工单位工作任务
☑	P01	施工单位任务	施工单位任务
☑	Q	Q-建设单位	建设单位工作任务
☑	Q01	建设单位任务	建设单位任务
☑	R	R-设计单位	设计单位工作任务
☑	R01	设计单位工作	设计单位工作任务
☑	S	S-勘察单位	有关勘察单位需要的功能
☑	S01	勘察单位工作	勘察单位工作功能入口
☑	T	T-新闻公告管理	新闻公告管理功能
☑	T01	新闻公告	新闻公告管理功能入口
☑	U	U-系统待设功能	系统待设功能
☑	U01	系统待设功能	系统待设功能入口
☑	U02	新增加功能	新增加功能

◉ 全选　◉ 全清　⬆ 确认　功能数量:134

图 7-15　角色功能设置视图运行界面
（图中内容是功能目录部分内容的截图）

往来单位子功能完成对往来单位信息的管理任务，其工作有往来单位记录检索列表显示、新增记录、记录编辑、记录详细内容显示、记录删除等常规的 CRUD 操作。功能实现的所有方法定义在控制器"LUnitListController"类中，其中"Index"是记录检索列表显示方法，具体代码内容如下：

```
//GET:/LUnitList/
public virtual ActionResult Index(FormCollection fc)
{
    string code1 ="";
```

```
        string code2 = "";
        int prows = 20;
        if( fc. Count > 0 )
        {
            apps. SetSession( "code1", fc[ "code1" ] );
            apps. SetSession( "code2", fc[ "code2" ] );
            apps. SetSession( "rowsperpage", fc[ "rowsperpage" ] );
        }
        code1 = apps. GetSession( "code1" );
        code2 = apps. GetSession( "code2" );
        prows = String. IsNullOrEmpty( apps. GetSession( "rowsperpage" ) ) ?
            20 : Convert. ToInt32( apps. GetSession( "rowsperpage" ) );
        ViewBag. code1 = code1;
        ViewBag. code2 = code2;
        ViewBag. rowsperpage = prows;
        var rd = db. LUnitLists. Where( u = > u. unitcode. Contains( code1 ) &&
            u. unitname. Contains( code2 ) );
        return View( rd. ToList( ) );
}
```

方法通过 "FormCollection fc" 接受来自视图提交的变量集合，并根据变量值完成数据记录检索，通过 "return View(rd. ToList( ));" 传递到视图进行显示。

"Index" 方法对应的视图文件名称是 "Index. cshtml"，具体代码内容如下：

```
@ model IEnumerable < psjlmvc4. Models. LUnitList >
@ {
    ViewBag. Title = "往来单位管理";
    string pkey = "";
    int i = 1;

    string code1 = ViewBag. code1;
    string code2 = ViewBag. code2;
    var grid = new WebGrid( source : Model,
        columnNames : new string[ ] { "unitcode", "unitname" },
        defaultSort : "unitcode",
        rowsPerPage : ( int ) ViewBag. rowsperpage,
        canPage : true,
        canSort : true,
        ajaxUpdateContainerId : "glist",
```

```
            fieldNamePrefix:"grid_",
            selectionFieldName:"selectedRow");
}
<table id=glist class="table"style="border:none;width:100%;">
    <caption><h2>@ViewBag.Title</h2></caption>
    <tr>
        <td style="border:none;width:70%;padding:0;">
            @grid.GetHtml(
                tableStyle:"tableStyle",
                headerStyle:"headerStyle",
                alternatingRowStyle:"",
                caption:"往来单位信息列表",
                selectedRowStyle:"SelectedRowStyle",
                footerStyle:"footerStyle",
                rowStyle:"",
                columns:grid.Columns(
                    grid.Column(header:"序号",format:item=>grid.PageIndex*grid.RowsPerPage+(i++),style:"selectColumnStyle"),
                    grid.Column("unitcode","单位编号",style:"fcode"),
                    grid.Column("unitname","单位名称"),
                    grid.Column(header:"选择",format:@<text>@item.GetSelectLink()</text>,style:"noprint")),
                mode:WebGridPagerModes.All,
                firstText:"第一页",
                previousText:"上一页",
                nextText:"下一页",
                lastText:"最后页",
                numericLinksCount:5,
                displayHeader:true,
                emptyRowCellValue:"",
                fillEmptyRows:false,
                exclusions:new string[]{"unitcode","unitname"},
                htmlAttributes:new{id="grouplist"})
            <div style="padding:3px 20px 3px 20px;">
                记录数:@Html.Encode(grid.TotalRowCount)
                总页数:@Html.Encode(grid.PageCount)
                当前页:@Html.Encode(grid.PageIndex+1)
                当前行:@Html.Encode(grid.PageIndex*grid.RowsPerPage+grid.SelectedIndex+1)
            </div>
        </td>
```

```
              <td id = "opp-div"style = "vertical-align:top;text-align:left;border-left:5px double #0094ff;">
                  <h2>操作选择</h2>
                  <hr/>
                  @using(Html.BeginForm())
                  {
                      @Html.Encode("单位编号")
                      @Html.TextBox("code1",(string)ViewBag.code1,new{style = "width:200px"})
                      <br/>
                      @Html.Encode("单位名称")
                      @Html.TextBox("code2",(string)ViewBag.code2,new{style = "width:200px"})
                      <br/>
                      @Html.Encode("每页行数")
                      @Html.TextBox("rowsperpage",(int)ViewBag.rowsperpage,new{style = "width:100px"})
                      <hr/>
                      <p>
                          <input type = "submit"value = "查询"class = "btn btn-primary"/>
                          @Html.ActionLink("新增","Create",null,new{@class = "btn btn-primary"})
                          @Html.ActionLink("返回","Index","Home",new{Area = ""},new{@class = "btn btn-primary"})
                      </p>
                  }
                  <hr/>
                  @if(grid.HasSelection)
                  {
                      WebGridRow wr = new WebGridRow(grid,grid.SelectedRow,grid.SelectedIndex);
                      pkey = Convert.ToString(wr.ElementAt(0));
                      <p>
                          @Html.ActionLink("编辑","Edit",new{id = pkey},new{@class = "btn btn-primary"})
                          @Html.ActionLink("详细","Details",new{id = pkey},new{@class = "btn btn-primary"})
                          @Html.ActionLink("删除","Delete",new{id = pkey},new{@class = "btn btn-primary"})
                      </p>
```

```
            <hr/>
            @Html.Encode(wr.ElementAt(0)) <br/>
            @Html.Encode(wr.ElementAt(1))
        }
        </td>
    </tr>
</table>
```

视图中记录显示的方法使用了"WebGrid"帮助器，视图运行后的界面如图 7-16 所示。

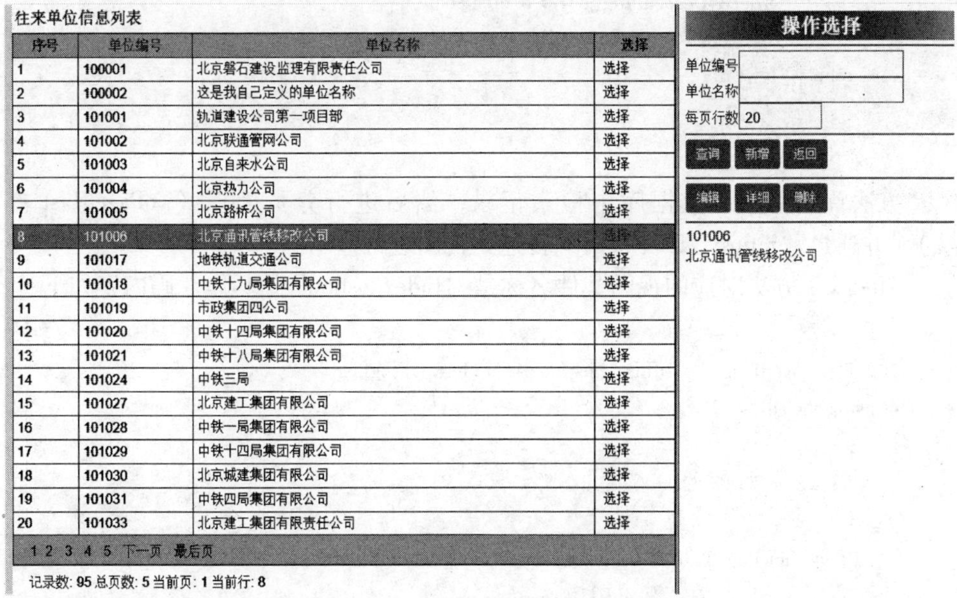

图 7-16　往来单位功能视图运行界面

界面内容由记录显示和操作选择两个部分组成，分左右排列，并且"编辑"、"详细"、"删除"三个操作不是和记录同步显示，而是当选中记录时才在右边显示，这是和前边的显示方式不同的地方。

7.4.4　O-其他辅助信息管理

其他辅助信息管理是为补充基础数据以外的数据而特别设置的功能，目前其包含"发言内容类别"管理和"用户发言"管理两个子功能。其中"发言内容类别"是为了对用户的发言根据内容性质进行分类管理的类别信息。实现该功能所需要的全部文件资源存储于区域"Areas/OArea"中。现以"用户发言"管理功能为例说明其实现的方法和过程。

用户发言功能完成用户发言内容的显示和新增加用户发言任务。功能实现的方法是"Index",定义在控制器"AFFileController"类中,具体代码内容如下:

```csharp
//GET:/OArea/BlogList/
public ActionResult Index( int page = 1)
{
    ViewBag.blogtypeid = new SelectList( db.OBlogTypes,"blogtypeid",
            "blogtypename","发言内容类别");
    var pagesize = 5;
    var rd = db.OBlogLists.
        OrderByDescending( o = > o.createdate).
        ToPagedList( page,pagesize);
    return View(rd);
}
```

方法首先完成从数据库中检索记录,然后进行分页处理(ToPagedList方法),并通过"return View(rd);"传递到视图进行显示。

"Index"方法对应的视图文件名称是"Index.cshtml",具体代码内容如下:

```cshtml
@model PagedList.IPagedList < psjlmvc4.Models.OBlogList >
@using PagedList.Mvc
@{
    ViewBag.Title ="用户发言记录";
}
< h2 > @ViewBag.Title </h2 >
< table id ="table"style ="width:100%;" >
@foreach( var item in Model) {
    < tr style ="background - color:#c4cbd3;font - size:1.2em;" >
        < td >
            @Html.DisplayFor( modelItem = > item.LUserStaff.fullname)
        </td >
        < td >
            @Html.DisplayFor( modelItem = > item.OBlogType.blogtypename)
        </td >
        < td >
            @Html.DisplayFor( modelItem = > item.createdate)
        </td >
```

```
            </tr>
            <tr>
                <td>
                    <img alt="ttt" src="@Url.Action("ImageFile","LUserStaff",new{Area="LArea",uid=item.userid})" width="100" height="100"/>
                </td>
                <td colspan="2" style="text-align:left;vertical-align:top;text-indent:2em;">
                    @Html.DisplayFor(modelItem=>item.blogcontent)
                </td>
            </tr>
}
</table>
<hr/>
@Html.PagedListPager(Model,page=>Url.Action("Index",new{page}))
<div style="background:linear-gradient(to top,#1e5799,#7db9e8 100%);height:30px;margin-bottom:5px;"></div>
@using(Html.BeginForm("Create","BlogList",FormMethod.Post))
{
    <div style="margin-bottom:5px;">
        @Html.Label("选择内容类别")
        @Html.DropDownList("blogtypeid")
    </div>
    <div>
        @Html.TextArea("blogcontent",new{rows=10,style="width:70%;"})
    </div>
    <p style="margin-top:10px;">
        <input type="submit" value="提交" class="btn btn-primary"/>
        @Html.ActionLink("返回","Index","Home",new{Area=""},new{@class="btn btn-primary"})
    </p>
}
```

  分页使用的类是"PagedList",在控制器和视图中都需要引入(using),在视图上显示分页的方法是通过"@Html.PagedListPager(Model,page=>Url.Action("Index",new{page}))"帮助器实现的。视图运行后的界面如图7-17所示。

  上部分是已发言内容显示,中间是分页操作,下部分是当前用户填写新的发言,并通过"提交"命令完成发言内容的存储。完成"提交"功能的方法"Create"的代码内容如下:

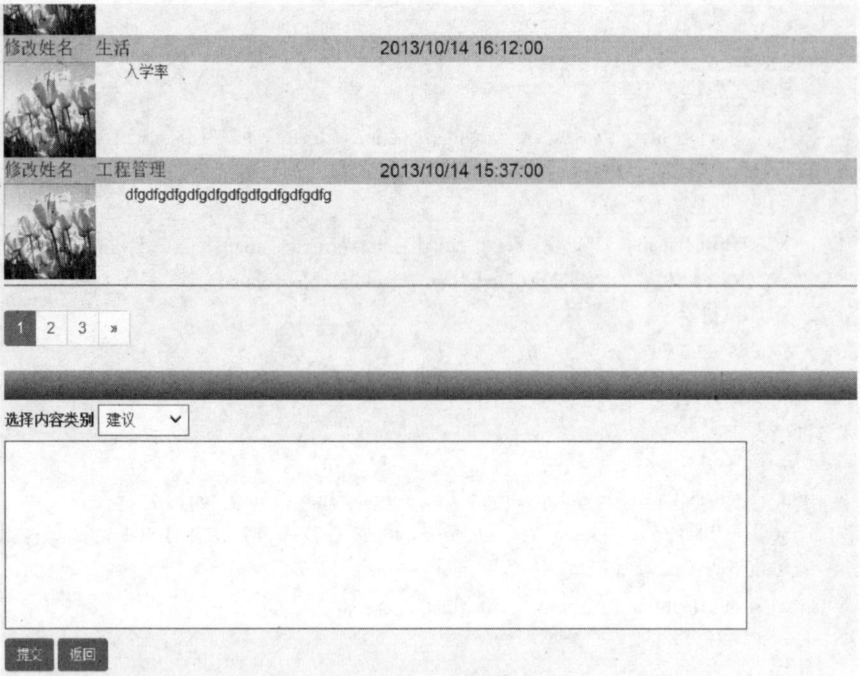

图 7-17　用户发言功能视图运行界面

```
[HttpPost]
public ActionResult Create(string blogtypeid = null, string blogcontent = null)
{
    OBlogList rd = new OBlogList();
    rd.userid = apps.GetSession("userid");
    rd.createdate = DateTime.Now;
    rd.blogtypeid = blogtypeid;
    rd.blogcontent = blogcontent;
    db.OBlogLists.Add(rd);
    db.SaveChanges();
    return RedirectToAction("Index");
}
```

本章小结

　　本章选取了系统主页功能模块、通用功能模块、工程管理功能模块和其他功能模块依次说明各功能设计实现的方法及代码内容。

参 考 文 献

[1] DBJ 01-41—2002，北京市地方性标准《建设工程监理规程》[S]．
[2] 蒋金楠．ASP.NET MVC 4 框架揭秘 [M]．北京：电子工业出版社，2013 年 1 月．
[3] 加洛韦．ASP.NET MVC 4 高级编程 [M]．北京：清华大学出版社，2013 年 8 月．
[4] 李智慧．大型网站技术架构：核心原理与案例分析 [M]．北京：电子工业出版社，2013 年 9 月．
[5] Jeffrey Palermo，Ben Scheirman Jimmy，Boggard．ASP.NET MVC 实战 [M]．北京：人民邮电出版社，2010 年 12 月．
[6] 蒋金楠．ASP NET MVC 5 框架揭秘 [M]．北京：电子工业出版社，2014 年 8 月．
[7] Andrew Troelsen．精通 C#（第 6 版）[M]．北京：人民邮电出版社，2013 年 7 月．
[8] 桂素伟．7 天精通 C#教程第 6 讲结构和枚举 [M]．北京：机械工业出版社，2010 年 5 月．
[9] 陈永强，李茜．SQL Server 2005 + PowerBuilder 11 高级开发指南 [M]．北京：清华大学出版社，2008 年 3 月．
[10] 张子阳，余昭辉，王波．C#揭秘 [M]．北京：人民邮电出版社，2010 年 4 月．
[11] 明日科技．SQL Server 从入门到精通 [M]．北京：清华大学出版社，2012 年 9 月．
[12] Aofengdaxia．Asp.net MVC 中 Ajax 的使用 [OL]．http：//blog.csdn.net/aofengdaxia/article/details/6880020，2011 年 10 月．
[13] Jingmeifeng．Razor 视图引擎—基础语法 [OL]．http：//blog.csdn.net/jingmeifeng/article/details/7788383，2012 年 7 月．
[14] 刘林，王新．管理信息系统 [M]．北京：科学出版社，2006 年 7 月．
[15] 王新．基于 DW 技术的管理信息系统分析设计实践 [M]．北京：对外经济贸易大学出版社，2013 年 3 月．